NEW CONCEPTS IN
WATER PURIFICATION

NEW CONCEPTS IN WATER PURIFICATION

Gordon L. Culp
Culp, Wesner, and Culp
Clean Water Consultants
El Dorado Hills, California

and

Russell L. Culp
Culp, Wesner, Culp
Clean Water Consultants
El Dorado Hills, California

VNR **Van Nostrand Reinhold Company**
New York / Cincinnati / Toronto / London / Melbourne

Van Nostrand Reinhold Company Regional Offices:
New York   Cincinnati   Chicago   Milbrae   Dallas
Van Nostrand Reinhold Company International Offices:
London   Toronto   Melbourne

Manufactured in the United States of America
Published by Van Nostrand Reinhold Company
450 West 33rd Street, New York, N.Y. 10001
Published simultaneously in Canada by Van Nostrand Reinhold Ltd.
15 14 13 12 11 10 9 8 7 6 5 4 3 2 1

Library of Congress Cataloging in Publication Data
Culp, Gordon L
  New concepts in water purification.

  (Van Nostrand Reinhold environmental engineering
series)
Includes bibliographies.
  1. Water-Purification. I. Culp, Russell L.,
1916-   joint author. II. Title.
TD430.C84 628.1'6 74-1067
ISBN 0-442-21781-1

Van Nostrand Reinhold Environmental Engineering Series

NEW CONCEPTS IN WATER PURIFICATION, by Gordon L. Culp and Russell L. Culp

ADVANCED WASTEWATER TREATMENT, by Russell L. Culp and Gordon L. Culp

ARCHITECTURAL INTERIOR SYSTEMS — Lighting, Air Conditioning, Acoustics, John E. Flynn and Arthur W. Segil

SOLID WASTE MANAGEMENT, by D. Joseph Haggerty, Joseph L. Pavoni and John E. Heer, Jr.

THERMAL INSULATION, by John F. Malloy

AIR POLLUTION AND INDUSTRY, edited by Richard D. Ross

INDUSTRIAL WASTE DISPOSAL, edited by Richard D. Ross

MICROBIAL CONTAMINATION CONTROL FACILITIES, by Robert S. Runkle and G. Briggs Phillips

SOUND, NOISE, AND VIBRATION CONTROL, by Lyle F. Yerges

Van Nostrand Reinhold Environmental Engineering Series

THE VAN NOSTRAND REINHOLD ENVIRONMENTAL ENGINEER-
ING SERIES is dedicated to the presentation of current and vital information relative to the engineering aspects of controlling man's physical environment. Systems and subsystems available to exercise control
of both the indoor and outdoor environment continue to become more
sophisticated and to involve a number of engineering disciplines. The
aim of the series is to provide books which, though often concerned
with the life cycle — design, installation, and operation and maintenance — of a specific system or subsystem, are complementary when
viewed in their relationship to the total environment.

Books in the Van Nostrand Reinhold Environmental Engineering
Series include ones concerned with the engineering of mechanical systems designed (1) to control the environment within structures, including those in which manufacturing processes are carried out, (2) to
control the exterior environment through control of waste products
expelled by inhabitants of structures and from manufacturing processes. The series will include books on heating, air conditioning and ventilation, control of air and water pollution, control of the acoustic environment, sanitary engineering and waste disposal, illumination, and
piping systems for transporting media of all kinds.

Preface

Water purification in the United States and in many other countries is now confronted with a myriad of new and difficult problems. Water supply sources are receiving greatly increased pollutional loads of domestic wastes and many new complex chemical wastes from industrial and agricultural operations. The presence of heavy metals, pesticides, herbicides, and the like in drinking water requires application of processes not widely used in the past in conventional water treatment plants.

It is the purpose of this book to describe such treatment techniques which are relatively new to the water treatment field and which have not been adequately described in other available textbooks. The descriptions are intended to emphasize practical and not theoretical aspects. Some of these new techniques may be applied to existing treatment facilities to increase their efficiency and capacity at a fraction of the cost required to construct new facilities to achieve the same goals. Thus, such improvements are more likely to be achieved and will provide the benefits of improved water quality with increased quantities available.

A conscious effort has been made to eliminate extensive recitation of material readily available in other texts. Thus, the reader may desire to supplement the material contained herein with the appropriately referenced material. The goal of the authors is to present material not available elsewhere under one cover on such topics as techniques for : practical application of shallow depth sedimentation including the first publication of detailed design criteria for various applications, increasing filter rates with concurrent gains in effluent quality and margins of safety, continuous monitoring and automatic control of the coagulation process, disposal or reclamation of chemical sludges resulting from water treatment, assuring adequate disinfection, removal of heavy metals from contaminated supplies, the integration of granular carbon into filtration systems to remove organic compounds, and the proper treatment of supplies containing large quantities of wastewaters. Examples of full-scale use of the techniques described are presented throughout the text to provide the reader with support material for use of the techniques in other locales. It is hoped such examples will also enable the reader to readily contact these existing facilities for further and updated information.

These new concepts in water purification can be applied to remedy a wide variety of conditions which are of current public concern. They can be used to treat water supplies which are receiving greatly increased pollutional loads of domestic wastes and new complex chemical wastes from industrial and agricultural operations. They can be used to remove pesticides, herbicides, some heavy metals, and other substances which are objectionable even when present only in trace amounts. They can remove taste and odor from water and eliminate the hazards involved when unpalatable tap water drives consumers to the purchase of bottled water. Bottled water may taste better but often is of very questionable bacteriological quality, and always costs much more. They can be used to treat runoff from watersheds which were once considered to be "protected," but which are now open to the pollutional threats posed by the widespread use of trail bikes, four-wheel drive vehicles, and snowmobiles which enable people to reach formerly inaccessible areas.

It is the hope of the authors that the material presented in this book will be of interest and benefit to engineers, chemists, bacteriologists, biologists, treatment plant operators, utility managers, administrators, and others faced with the daily tasks of furnishing or assuring supplies of high quality water to the public, and that the book will prove to be valuable as a reference for teachers and students who are interested in studying new methods for enhancement of water quality.

The quest for pure water can benefit the life and health of everyone. In the early 1900s waterborne diseases were rampant, and it was only through the immediate and widespread application of the then new processes of rapid-sand filtration and chlorination to purification of public water supplies by sanitary engineers in the United States that epidemics of these diseases were brought under control. The resulting improvements in public health were great and quite dramatic, and represent one of the major health advances in all history. Since that time, the general safety and high quality of public water supplies in this country have been a continuing source of satisfaction and pride, not only to people in the water works industry, but to most other Americans as well.

Early water treatment methods were developed almost entirely as an art rather than as a science. Today, water research is conducted along highly sophisticated scientific lines not only by engineers but also by many others engaged in a wide range of disciplines. There now exists a tremendous resource of scientific data on water and on solutions to water problems. While some of this information has been utilized in the development of new processes for improving water quality, a great deal more can be done in the art of practical application of the valuable new scientific knowledge now available. Many recently perfected water treatment processes must be included in plans for new plants and must be added to existing plants if the full potential benefits to the public from scientific and engineering progress in this area are to be realized. The means are now at hand to solve virtually any and all water quality problems.

The authors of this book have been fortunate to be involved in the development of several new and improved water treatment processes. They have also had the opportunity to incorporate these new methods into water treatment plant designs and to observe the results of full scale plant operation. These experiences are heavily drawn upon in this book. The work of many other investigators using other new methods are also referred to so as to present, insofar as possible, an accurate updating of currently available treatment methods.

Special thanks are given to CH2M/Hill, Consulting Engineers, and to the several equipment manufacturers who so generously supplied illustrations for the text. The encouragement of our families and colleagues throughout the preparation of the book is greatly appreciated.

GORDON L. CULP
RUSSELL L. CULP

Contents

NEW CONCEPTS IN
WATER PURIFICATION

1 Improving Water Quality In a Changing Environment

THE PRESENT SITUATION

Water Quality

The high quality of public water supplies in the United States has long been a source of local and national pride. Travelers drink water from the tap wherever they may be with no question of its safety. Conformance to the mandatory requirements of the U. S. Public Health Drinking Water Standards has proven to be an acceptable measure of water safety in the past. Restrictions on the quantity of water used and the occurrence of waterborne diseases in this country are not the serious problems that they are elsewhere in the world.

Despite this generally good record, water utilities are currently subjected to a great deal of criticism in the press, by the public, and in the Congress. There are many reasons for this. In many cases, water service is not nearly so good as it can and should be. Many community water systems are antiquated and do not meet present day standards of quality and service. There is lack of planning for future needs and a need for integrating an existing proliferation of small separate water systems. Local, state, and federal efforts at survelliance and monitoring, operator and manager training, and water research fall far short of today's needs. There is public demand for improved water quality.

1

Existing Deficiencies in Water Supplies

The findings of the U. S. Public Health Service Community Water Supply Study in 1969 were a severe shock to laymen and professionals alike. Approximately 17 percent of all public water supplies failed to meet one or more of the mandatory quality standards; 25 percent did not meet one or more of the recommended quality standards; more than 50 percent had major deficiencies in supply, storage, or distribution facilities; 90 percent failed to meet the bacteriological standards; and 90 percent had no cross connection control programs. Conditions were particularly bad in the very small systems. Many were substandard in almost every respect.

The State of Washington, Division of Health, has accumulated some statewide statistics in this regard which are of interest. Their findings are probably similar to conditions prevailing in other states. They found that somewhere between 20 and 50 Washington communities, including some of the largest cities, are in need of the addition of filtration treatment. The exact needs are not known due to the uncertainty which has arisen recently about the quality of supplies derived from so-called "protected watersheds." The easy public access to any and all areas including protected watersheds which has been provided by the advent of all-terrain vehicles, trail bikes, and snowmobiles has made it very difficult or impossible to assure safety of subsurface water supplies which have no facilities for filtration treatment.

The study reported that 55 supplies, including all of the largest cities in the state, have open distribution reservoirs for storage of finished water. In the judgment of the Washington Division of Health, open distribution reservoirs are not compatible with present good public health practice, and they must now be covered under state regulations.

In 1971, the needs for construction of water works improvements in the State of Washington to meet current needs were estimated to be more than $325 million, with future needs estimated at an additional $35 million per year through the 1970s. The results of bacteriological and chemical testing in Washington state corresponded with the nationwide findings of the Community Water Supply Study. Many city water supplies did not meet the Standards.

The Washington Division of Health is now requiring all cities to submit a comprehensive water system plan to identify present and 10 year future needs. It is anticipated that one third of the cities already have most of the necessary data for such a plan; one third will have a considerable amount of updating to do; and that the other one third have virtually none of the data required for advance planning and will have to start from scratch. Two of the major unmet planning needs are coordinated planning among neighboring water utilities, and minimization of the number of new systems being constructed.

With respect to water system operation, Washington found that there were needs for higher operator salaries and better employee benefits; for an adequate water department budget financed by rates which are kept current with system needs; and for stronger support of the local water department by local government officials and the public.

Statistics compiled by the State of Washington also reveal how little effort is being extended in the survelliance of water supplies by local, state, and federal regulatory agencies as compared to similar efforts in air and stream pollution control. Expenditures for air pollution control were 10 times that for water supply monitoring and that for stream pollution control 200 times greater. The Washington study concluded that the incidence of disease is no longer an acceptable measure of the adequacy of public water supply systems, and that there is a need to make this clear to the public, the press, and even to the water works industry.

In addition to the deficiencies already mentioned with respect to public health, there are also aesthetic considerations which deserve mention because of their importance to the consumer. Many water supplies contain sufficient iron or manganese to cause staining of laundry and plumbing fixtures, or even to building facings which are in the range of lawn sprinklers. Some waters are too hard to be satisfactory for many uses or have corrosive or scaling tendencies which adversely affect plumbing in homes or factories. One of the most common complaints about water quality is that of taste and odor. Bad tasting water drives many consumers to other sources for drinking water such as private wells, cisterns, or bottled water. More often than not, such alternate sources of drinking water are of very questionable bacteriological quality.

Trends in Raw Water Quality

Increased Water Withdrawals.   Surface water sources are subject to ever increasing withdrawals to supply a growing population, industry, and agriculture. Heavy withdrawals of water for consumptive use, such as irrigation, decrease stream flows available for downstream dilution of wastewaters. In the case of underground water supply sources, excessive withdrawals may have an adverse effect on the chemical quality of the supply, such as increasing the content of iron or manganese or of total dissolved solids.

Waste Discharges.   More importantly, ever increasing quantities of domestic, industrial, and agricultural wastewaters are being discharged directly or indirectly to water supply sources. Domestic sewage continues to add increasing quantities of bacteria, virus, algal nutrients, oxygen demand, suspended and dissolved solids, and taste and odor producing substances to water supply sources. Return irrigation waters in some locations are producing substantial increases in the total dissolved solids content of receiving streams and aquifers, and many of these substances such as sodium, sulfate, chloride, and others are not removed by conventional treatment methods or other economically feasible means. Serious changes are also produced by industries which are discharging a wide variety of new complex chemical pollutants to the waters of the nation. Recently, the Environmental Protection Agency and a private laboratory determined the presence of 34 identifiable chemicals in the water at the lower end of the Mississippi River — the drinking water supply for more than half the people in Louisiana. A recent listing of refractory organic wastes — some of which have been found in drinking water — is presented in Table 1-1 as an illustration of the complex waste materials which now may be entering our water supplies.

TABLE 1-1   REFRACTORY INDUSTRIAL WASTES (From *Environmental Science and Technology,* January, 1973)

Acetone	Dichloroethyl ether
Benzene[a]	Dinitrotoluene[a]
2-Benzothiozole	Ethylbenzene[a]
Borneol (bornyl alcohol)	Ethylene dichloride[a]
Bromobenzene	2-Ethylhexanol
Bromochlorobenzene	Guaiacol (methoxy phenol)
Bromophenylphenyl ether	Isoborneol
Butylbenzene	Isocyanic acid[a]
Camphor	Isopropylbenzene
Chlorobenzene	Methylbiphenyl
Chloroethyl ether	Methylchloride[a]
Chloroform	Nitrobenzene[a]
Chloromethyl ethyl ether	Styrene[a]
Chloronitrobenzene	Tetrachloroethylene
Chloropyridine	Trichloroethane
Dibromobenzene	Toluene[a]
Dichlorobenzene	Veratrole
	(1,2-dimethoxybenzene)

[a]In trace amounts, these compounds also have been found to impart taste and odor to drinking water supplies.

With adequate treatment of these increased wastewater flows by modern techniques prior to discharge, raw water quality will not suffer, and, in fact, may be substantially improved in many cases. Unfortunately, the construction of pollution control facilities is not keeping pace with increased flows. The rate of construction is so far behind today's needs that many receiving waters are being progressively degraded with the passing of each year. This lack of essential facilities is a matter of grave national concern.

Wastewater Treatment Deficiencies. In addition to the lack of adequate facilities, there are other problems in water pollution control. Many existing sewage and industrial waste treatment plants which discharge their treated waters to water supply sources are inadequately designed or are poorly operated and do not achieve their planned treatment efficiency. In the past, less attention has been given to plant reliability, backup, and standby facilities in the design of wastewater treatment plants than is commonly devoted to the design of power generation stations or water systems, for example. In addition, the operation of wastewater treatment plants is almost universally underfinanced and understaffed, and badly so. In the vast majority of plants, monitoring of effluent quality is inadequate or nonexistent, and control of plant operations and the efficiency of treatment are very poor. Prolonged complete bypassing of wastewater treatment plants has been quite common, and it is not unusual for the entire contents of a sour sludge digestor to be dumped to a receiving watercourse when an opportunity is presented, such as high stream flow or the cover of night.

In view of this situation, it is obvious that there is a great opportunity to improve raw water quality and the safety of public water supplies through better operation of existing wastewater treatment plants, the expansion of plants to handle current sewage flows and provide more reliable operation, the addition of facilities for advanced waste treatment, adequate financing of operations, the training of plant operators, and better survelliance of plant efficiencies. There is a great deal of talk about a nationwide attack on all these fronts, and many political promises of progress in this area. However, water purveyors would be well-advised to plan their treatment needs and design their plants on the basis of what is actually being accomplished in stream cleanup. The sad fact is that very little improvement in raw water quality is occurring, and the prudent course for advance planning is to anticipate even further deterioration of raw water quality, and to plan, construct, and operate water purification facilities accordingly.

Some Special Problems. The interreaction of many water quality control problems and proposed solutions can create even more complex or new problems for the water treatment industry. An illustrative example is offered by the seemingly separate problems of control of heavy metal discharges and the reduction of phosphorus discharges to watercourses.

Because of the high toxicity of mercury, disclosures of the discharge of substantial quantities of mercury in wastes from paper mills to the water environment, and its fate there, aroused a great deal of interest both on the part of water purveyors and the general public. It also raised the question of the possible undetected presence of other heavy metals in water supplies.

The possible presence of trace quantities of heavy metals in water supplies would be affected by the use of NTA (nirilotriacetic acid) as a substitute or supplement for phosphorus compounds in the manufacture of synthetic detergent washing products. NTA was, at one recent point in time, considered the most promising substitute for phosphorus in detergents and the detergent industry was tooling up to produce massive quantities of NTA. NTA is a chelating agent which has the property of increasing the solubility of heavy metal ions in water. NTA also can prevent the formation of insoluble salts of metal ions. Thus, NTA could cause serious interference with wastewater treatment and water purification. By sequestering heavy metal ions, some of which are quite toxic, the metal ions may escape removal by coagulation, sedimentation, filtration, or other processes which would otherwise be effective in their elimination. As a result, potentially toxic ions could pass into receiving streams. If present in sufficient concentration in surface water supply sources, NTA could similarly reduce the efficiency of water purification plants. While there is a question, but no absolute proof at this time, that the ingestion of trace amounts of NTA itself over a long period of time would have any ill effects on health, the potential harmful effects of some heavy toxic metals which may slip by water treatment due to chelation by NTA were certainly well-known. At one point, prior to the U. S. Environmental Protection Agency restricting its use in detergents, about 100 million lb per year of NTA were going into detergents, and additional production capacity of about 300 million lb per year was under construction. This use of NTA was stimulated by public and political pressures to remove phosphorus from detergents. Phosphorus stimulates algal growths, and under some circumstances produces nuisance conditions. However, there are no ill effects, and very efficient and inexpensive methods have been developed for removing phosphorus from wastewater. Several advanced wastewater treatment (AWT) plants now routinely remove 99+ percent of phosphorus from wastewater. Only about one half the phosphorus in wastewater originates from the use of detergents, the other half has always been present and will continue to be present even if all phosphorus is eliminated from detergent products. Thus, the quantities of phosphorus remaining in wastewaters even if phosphorus were removed from detergents would still be adequate to stimulate algae problems. Further, doubling the quantity of phosphorus to be removed by AWT only very slightly increases the total chemical requirements and costs for removal of phosphorus. This is not to say that phosphorus should or should not be removed from detergent formulations, but only to point out that any trade-off of NTA for phosphorus is a poor bargain indeed. It appears that manufacturers were stampeded into an unwise change which could have produced substantially increased environmental hazards and major problems for the water supply industry.

In 1966, the Denver Board of Water Commissioners conducted a study of the Denver water supply to characterize the chemical composition of the water supplies particularly with respect to minor elements. Of the 25 trace elements tested for, the following 13 were found: aluminum, barium, boron, chromium, copper, iron, lithium, manganese, molybdenum, rubidium, strontium, titanium, and zinc. The other 12 elements which were not found were: beryllium, bismuth, cadmium, cobalt, gallium, germanium, lead, nickel, silver, tin, vanadium, and zirconium. One unexpected result was the finding of rather high (maximum 260 mg/l) concentration of molybdenum in two of the supply reservoirs by virtue of their receiving runoff from areas in which there are extensive molybdenum mining operation. Molybdenum was not removed by the process employed in the treatment plants.

As a result of the discharge of increased quantities and new types of pollutants in wastewaters, water purveyors are faced with a myriad of new problems in plant design and operation. Many people are concerned that the factor of safety which public water supplies have enjoyed in the past is being reduced to a dangerous extent by current trends in the degradation of raw water quality. Fortunately, many professionals in the water works field are seriously concerned with these problems and are actively working within the limits of their resources to find practical solutions to them. There also is considerable public support for improving present conditions.

IMPROVING WATER QUALITY

In the face of the probable continued degradation of raw water quality, the prospects for improvement of finished water quality may not appear bright. Fortunately, this is not the case. Despite the difficulties posed by increased levels of pollution, it is still possible and feasible to substantially improve water quality. Much of this potential improvement can be realized by judicious application of conventional purification processes, but even further gains can be had by utilizing new developments in water treatment such as mixed-media filtration combined with the use of alum or polymer as a filter aid, proper instantaneous and thorough mixing of chlorine and the water being treated to provide better disinfection, the use of granular activated carbon for removal of trace organic compounds and for complete color removal and taste and odor control, practical applications of shallow depth sedimentation (tube settling), the use of polymers to improve coagulation, flocculation, and settling, the provision of methods for recycling or satisfactory disposal of sludges, the use of methods for continuously monitoring and controlling water treatment processes, and other methods. These possibilities are to be explored in detail in the sections which follow.

Objectives of Water Treatment

The primary objective of water treatment for public supply is to take water from the best available source and to subject it to processing which will assure that it is always safe for human consumption and is aesthetically acceptable to the consumer. For water to be safe for human consumption, it must be free of pathogenic organisms or other biological forms which may be harmful to health, and it should not contain concentrations of chemicals which may be physiologically harmful. To provide safe water, the treatment plant must be properly designed and skillfully operated. As already pointed out, the task of furnishing safe water in the future in many instances may be made more difficult because of the necessity of treating a raw water of poorer quality. The general requirements of an aesthetically acceptable water are that it be cool, clear, colorless, odorless, and pleasant to the taste; also it should not stain, form scale, or be corrosive. Treatment plants must be designed to produce water of uniformly good quality despite variations in raw water quality and plant throughput. Since the consumer is interested in the quality of the water at his tap rather than that at the treatment plant, precautions must be taken to preserve water quality in the distribution system and to control water quality from tests of tap water samples as well as plant samples. Many of the advantages of a high quality water supply are difficult to express in terms of exact economic return, but they are, without question, quite substantial. An ample supply of high quality water assures good public relations and favors industrial and community growth.

Standards

Currently, 1973, the minimum standards for water quality are the U. S. Public Health Service Drinking Water Standards of 1962.

Although these standards were promulgated for use in regulating the quality of water supplied to common carriers engaged in interstate commerce, they have become the standard for all public water supplies in the country through endorsement by the American Water Works Association and formal adoption by state health departments. The Standards set mandatory limits for certain chemical constituents and recommend concentrations for others including some radioactive elements. The bacteriological standards set limits for coliform organisms and prescribe methods for the collection and laboratory analysis of water samples including the frequency thereof.

TABLE 1-2   SUMMARY OF 1962 U. S. PUBLIC HEALTH SERVICE DRINKING WATER STANDARDS

MANDATORY REQUIREMENTS

Constituent	Limit (mg/l except as shown)	Range of Concentrations Found in National Survey[b] (mg/l except as shown)
Arsenic (As)	0.05	<0.03–0.10
Barium (Ba)	1.0	0 −1.55
Boron (B)	5.0	0 −3.28
Cadmium (Cd)	0.01	<0.2 −3.94
Chromium (hexavalent)	0.05	0 −0.079
Coliform organisms	1/100 ml	2000/100 ml
Cyanide (CN^{-2})	0.20	0 −0.008
Fluoride (F^-)	Varies[a] 0.8 to 1.7	<0.2 −4.40
Gross beta activity	1000 $\mu\mu c/l$	154 $\mu\mu c/l$
Lead (Pb)	0.05	0 −0.64
Selenium (Se)	0.01	0 −0.07
Silver (Ag)	0.05	0 −0.03

[a]Dependent on annual average of maximum daily air temperatures.
[b]1969 Community Water Supply Study by USPHS, 2595 distribution system samples.

The minimum number of water samples per month to be collected for bacteriological examination varies according to population served by the system from 2 per month for 2000 people or less to 100 per month for 100,000 people to 300 per month for 1,000,000 persons. The fact that the total number of bacteriological samples collected and analyzed is less than the minimum required for significant results is one of the most common reasons for failure to meet the Drinking Water Standards. Tables 1-2 and 1-3 summarize the current standards.

TABLE 1-3   SUMMARY OF 1962 U. S. PUBLIC HEALTH SERVICE DRINKING
WATER STANDARDS
RECOMMENDED REQUIREMENTS

Constituent	Recommended Limit (mg/l except as shown)	Range or Maximum Concentration Found In National Survey[a] (mg/l except as shown)
Alkyl benzene sulfonate (ABS)	0.5	0 – 0.41
Arsenic (As)	0.01	0 – 3.28
Chloride (Cl⁻)	250	<1.0 –1950
Color	15 units	49 units
Copper (Cu)	1.0	0 – 8.35
Carbon chloroform extract (CCE)	0.200	0.008– 0.56
Cyanide (CN⁻²)	0.01	0 – 0.008
Fluoride (F⁻)	Varies 0.8 to 1.7	<0.2 – 4.40
Iron (Fe)	0.3	26.0
Manganese (Mn)	0.05	1.32
Nitrate (No₃)	45	0.1 – 127.0
Phenols	0.001	
Radium-226	3 pc/l	0 – 135.9 pc/l
Strontium-90	10 pc/l	0 – 2.0 pc/l
Sulfate (SO₄)	250	<1 – 770
Total dissolved solids (TDS)	500	2760
Zinc (Zn)	5	0 – 13.0
Turbidity		
Chlorination treatment only	5 JTU	53 JTU
Water treatment plants	1 JTU	
Odor, threshold number	3	

[a]1969 Community Water Supply Study by USPHS, 2595 distribution system samples.

Goals

In 1968, the American Water Works Association adopted a statement of water quality goals which are more exacting than the existing USPHS Drinking Water Standards with respect to aesthetic qualities, yet they are practical objectives. They are attainable by correct application of known treatment processes and methods.

The goals do not include certain toxic substances which are adequately covered by the Standards, as follows:

Lead	Silver	Radium	Organic
Barium	Selenium	Strontium	phosphorus
Fluoride	Cadmium	Phenolic	Chlorinated
Arsenic	Chromium	compounds	hydrocarbons
Cyanide	Nitrates		Boron
			Uranyl ion

Table 1-4 summarizes the AWWA water quality goals.

TABLE 1-4   SUMMARY OF AWWA POTABLE WATER QUALITY GOALS

Characteristic	Goal
Physical Factors	
Turbidity	Less than 0.1 JTU
Nonfilterable residue	Less than 0.1 mg/l
Macroscopic and nuisance organisms	No such organisms
Color	Less than 3 units
Odor	No odor
Taste	No objectionable taste
Chemical Factors in mg/l	
Aluminum (Al)	Less than 0.05
Iron (Fe)	Less than 0.05
Manganese (Mn)	Less than 0.01
Copper (Cu)	Less than 0.2
Zinc (Zn)	Less than 1.0
Filterable residue	Less than 200
Carbon chloroform extract (CCE)	Less than 0.04
Carbon alcohol extract (CAE)	Less than 0.10
Methylene-blue-active substances (MBAS)	Less than 0.20
Radiologic Factors	
Gross beta activity	Less than 100 pc/l
Bacteriologic Factors	
Coliform organisms by multiple-fermentation or membrane filter techniques	No coliform organisms
Corrosion and Scaling Factors	
Hardness (as $CaCO_3$)	80 to 100 mg/l
Alkalinity change as $CaCO_3$ in distribution system, or after 12 hrs. at 130°F in closed plastic bottle, followed by filtration	Less than ± 1 mg/l
Coupon tests(90 day tests):	
Incrustation of stainless steel	Not more than 0.05 mg/cm^2
Corrosion of galvanized iron	Not more than 5 mg/cm^2

Recent Developments

The health and aesthetic significance of a number of trace elements and compounds have been questioned and have received consideration following adoption of the 1962 Standards. The U. S. Public Health Service has set unofficial limits for some of these substances while others have not been set and are still under consideration. It is possible that limiting concentrations of part or all of these substances will be included in the next revision of the Drinking Water Standards. Table 1-5 summarizes these items.

**TABLE 1-5 TENTATIVE SUGGESTED LIMITS OF CERTAIN TRACE ELE-
MENTS AND COMPOUNDS NOT INCLUDED IN 1962 USPHS
DRINKING WATER STANDARDS**[a]

Ingredient	Suggested Limit, mg/l	
Chemical Factors		
Antimony (Sb)	0.05	
Berryllium (Be)	None set	
Bismuth (Bi)	None set	
Boron (B)	1.0	
Molybdenum (Mo)	None set	
Mercury (Hg)	0.005	
Nickel (Ni)	None set	
Sodium (Na), limit for persons with certain diseases, only	20.0	
Uryanyl ion (UO₂)	5.0	
Pesticides		
Aldrin	0.017	
Chlordane	0.003	
DDT	0.042	
Dieldrin	0.017	
Endrin	0.001	
Heptachlor	0.018	
Heptachlor epoxide	0.018	
Lindane	0.056	
Methoxychlor	0.035	
Toxaphene	0.005	
Organic phosphorous plus carbamates, such as Parathion, Malathion, Carbaryl, and others	0.1	
Herbicides		
Algicide	1.0	
Copper key-El	1.0	(Cu)
Cuprose	1.0	(Cu)
Cutrine	1.0	(Cu)
Dipuot	0.2	
Dalapon	0.1	
2.4-D	1.0	
2,4,5-T should not be used		
Micro-gard PR	0.0	
Mogul algicide AG-470	1.0	
Radapon	0.1	
Silvex (2,4,5-TP)	0.2	
Tordon 10K & 101	0.0	
Twink	0.2	
Ureabor	1.0	(as B)

[a]After Floyd B. Taylor, "Trace Elements and Compounds in Waters," *Journal, AWWA*, 728 (1971).

Quality Concepts and Rationale

In assessing the significance of the failure of a water to meet the Standards or goals for high quality water in certain respects, it is helpful to know some of the reasons or background for the limitations set on certain constituents. Such information is also useful in evaluating the extent of the hazards involved in use of water not conforming to established criteria. Brief explanations of some of this background data are given below.

Bacteriological Quality. The most sensitive and valuable currently available tool for determining the effectiveness of water treatment processes in removing bacterial contamination is the coliform test. Coliform bacteria inhabit the intestinal tract of man and other warm-blooded animals. The predominance in sewage of coliform bacteria makes this organism a very sensitive indicator of pollution. Water is an unfavorable environment for bacteria and those that find their way into water gradually die off. Coliforms along with other bacteria are also quite readily removed from water by conventional water purification processes. The common intestinal bacterial pathogens are at least as susceptible to the natural and artificial purification processes to which water is subjected as are the more common coliform bacteria. Therefore, the coliform group is a good indicator of pollution. Detection of coliforms in small quantities of water (50 ml or less) suggests pollution, whereas their absence is considered an indication of probable safety. Despite intensive efforts over the years to find a better test for bacteriological quality, the coliform test is still the most useful.

By proper treatment, including chlorination as required, it is possible to produce a water which is free of coliform organisms. Further, by maintaining residual chlorine throughout the water distribution system, the bacteriological quality can be preserved until the water is drawn from consumer's taps. Many well-operated utilities have sampled daily for many years without detecting the presence of coliforms at any time in finished water.

Arsenic. The toxicity of arsenic in well-known, 100 mg usually causes severe poisoning. At low intake levels, a considerable portion is retained. Chronic arsenic poisoning is difficult to diagnose, and can be quite disabling prior to detection. Despite the widespread use of lead arsenate spray in orchards and grasslands, no reports have been made of contamination of water supplies by runoff from such watersheds.

Goldblatt, et al. [28] in 1963, reported the widespread occurrence of arsenic in well waters in Lane County (near the city of Eugene), Oregon. Concentrations of arsenic in some ground water of Lane

County was several times greater than the 0.05 mg/l limit, and the highest concentration found was 1.7 mg/l. In all 27 wells drawing ground water from an area underlain by the Fisher formation, the water was found to contain more than 0.05 mg/l of arsenic, and another 25 wells produced water containing from 0.02 to 0.05 mg/l of arsenic. This natural occurrence of arsenic in ground water is the only one reported to date in a well-populated area in the U. S. Although only two near-fatal cases of chronic arsenic poisoning were definitely attributable to the consumption of the arsenic-rich ground water in Lane County, there were undoubtedly many undetected cases of chronic toxicity in addition.

In 1943, Goudey disclosed that for several months a portion of the population served by the Los Angeles supply consumed water containing 1.0 mg/l of arsenic without known ill effects. Arsenic may also be present in hot springs or in certain industrial wastes. Arsenic can be removed from water in ion exchange equipment using activated alumina or bone char.

Barium.   Barium can have serious toxic effects on the heart, blood vessels, and nerves. The fatal dose for man is considered to be 550 to 600 mg. Barium salts are sometimes incorporated in rat poison. Barium occurs naturally in some mineral springs as the carbonate salt. Barium concentrations as high as 1.55 mg/l have been reported in public supplies. No studies have been made of the amounts of barium that may be tolerated in drinking water, but a rational basis for the water standard of 1.0 mg/l was derived from the threshold limit in air using well-established procedures.

Boron.   Boron ordinarily does not occur in most public supplies but has been reported in a few supplies in California and Italy. Goudey reported in 1939 that the boron content of the Los Angeles supply was 0.5 to 1.5 mg/l. Concentrations of 30 mg/l in drinking water may have some physiological effect. Ingestion of large amounts of boron can affect the central nervous system, and protracted ingestion may cause borism. Even in small amounts, boron is of importance in irrigation waters. An excess or deficiency can have serious effects on plant life.

Cadmium.   Cadmium is highly toxic. Concentrations of 13 to 15 mg/l in food have caused illness. Rat feeding studies also showed serious effects at levels of 15 mg/l cadmium. On the other hand, persons whose drinking water had an average cadmium content of 0.047 mg/l over a long period of time showed no ill effects. Cadmium can become a water contaminant through careless discharge of electroplating plant wastes, or from galvanizing operations in which cadmium is a contaminant.

Hexavalent Chromium.   When inhaled, chromium can cause cancer in man, but the results of ingestion are not known. The limit set by the Standards provides a safety factor. Trivalent chromium salts show none of the toxicity of the hexavalent form. The most likely sources of hexavalent chromium in water would be pollution by wastes from chromium-plating shops or through cross-connections to chromate-treated cooling water in air conditioning systems. In surface waters, toxic hexavalent chromium is modified to the harmless trivalent form, but caution is advisable until more is known about the toxicity of the hexavalent form.

Cyanide.   Usually, 50 to 60 mg of cyanide in a single dose is fatal. However, in doses of 10 mg or less, cyanide is converted to thiocyanate in the body. Usually the lethal effects occur only when the detoxifying mechanism is overwhelmed. Chlorination under neutral or alkaline conditions of water cyanide will reduce cyanide below the limit set in the Standards. The cyanogen chloride produced is about one twentieth as toxic as hydrogen cyanide. The drinking water limit provides a factor of safety of about 100 for humans. However, trout may be killed by concentrations of cyanide as low as 0.05 mg/l in water.

Fluoride.   Optimum concentrations (0.7 to 1.2 mg/l) of fluoride in drinking water consumed during childhood will prevent dental caries. Excessive concentrations (more than about 3.0 mg/l) can cause dental fluorosis, or mottled enamel, when consumed in water during the childhood years. When water containing 8 to 20 mg/l of fluoride is consumed over a long period of time bone changes may occur, and when 20 mg/l day of fluoride from all sources is consumed for 20 years or more, crippling fluorosis may result. Toxic efforts are not exhibited until much higher fluoride concentrations are reached. Based on drinking 2 liters of water per day, 115 mg/l would be a sublethal dose and 2000 mg/l (16,000 lb of sodium fluoride per million gallons) would be lethal.

The maximum concentration of fluoride in natural waters (principally well waters) usually does not exceed 5.0 mg/l. There are methods for removing fluoride from water or reducing the concentration to the optimum level, although the costs are more than for conventional treatment. Fluoride is often added to fluoride-deficient public water supplies to adjust the concentration to the optimum for control of dental caries.

Gross Beta Activity.   In the Standards, an upper limit of 1000 $\mu\mu$c/l of gross beta activity (in the absence of alpha emitters and strontium-90) has been set. This is basically a screening test. When this limit is exceeded, a complete analysis must be made to identify the specific radionuclides present to be sure that exposures are within the limits set by the Federal Radiation Protection Guide. The natural background in most areas is only about 10 pc/l.

Lead. It is common knowledge that lead is a serious cumulative poison. Lead taken into the body by either brief or prolonged exposures can seriously injure health or cause death. Common sources of lead intake are food, air, water, and tobacco smoke. The amount of lead in water that the body can tolerate depends upon the relatively uncontrollable intake from other sources. Persons can use water for a few weeks containing as much as 2 to 4 mg/l without ill effect, but use for more than 3 months may be harmful. Several weeks use at lead levels of 8 to 10 mg/l can product toxic effects. Consumption of water containing more than 15 mg/l of lead over a period of several weeks can be fatal. Concentration of 0.05 mg/l as set in the Standards as the maximum are physiologically safe in water. Lead in soft water is highly toxic to some small fish at concentrations as low as 0.10 mg/l.

Selenium. Selenium is highly toxic to many animals and fish. Its effects may be permanent. Selenium is also a potential carcinogen. In certain areas where the soil contains appreciable quantities of selenium, it may also be present in water supplies, principally in shallow wells.

Silver. Ingestion of sufficient silver (about 1 g) can produce a permanent irreversible blue-gray discoloration of the skin, eyes, and mucous membranes which is unsightly and disturbing. Assuming that all silver ingested is deposited in a single body integument, a 10 μg/l could be ingested for a lifetime or 50 μg/l of silver could be ingested for about 27 years without exceeding silver deposition of l g.

Alkyl Benzene Sulfonate (ABS). The standard of 0.5 mg/l gives a safety factor of about 15,000 times so far as oral toxicity is concerned, but concentrations higher than the Standard may cause the water to foam or to have a bad taste. With the advent and predominant use of biodegradable detergents, ABS is no longer the problem in water supplies that it was in the past. The test for all MBAS (methylene blue active substances) has replaced the procedure for ABS in most water laboratories.

Methylene Blue Active Substances (MBAS). The AWWA Quality Goals are of more recent issue than the Standards, and the ABS Standard has been replaced with an MBAS goal of 0.2 mg/l. The change of designation is necessary because of changes in composition of the new detergents. The MBAS analytic technique used determines alkyl sulfates and related materials that react with methylene blue as well as ABS.

Chloride, Sulfate and Dissolved Solids. These substances affect water quality with regard to its taste and its laxative properties. The taste threshold and laxative effects of various salts in water vary with individual sensitivity, the presence of other substances, and individual acclimatization to the use of a particular water. Therefore, the acceptable concentration is a range of values rather than a specific concentration for each salt. Various studies support reasonably well the Standards of 250 mg/l for chlorides and sulfates, and 500 mg/l for total solids. However, it is generally recognized that waters with concentrations of these salts exceeding the Standards are widely used without ill effect. People apparently become acclimated to the use of these high solids waters in a relatively short time. A more rational approach than the use of TDS as a controlling standard is the use of standards on individual ions.

Turbidity, Color and Odor. The requirements of the Standards in regard to these physical characteristics of water are quite easily met by treatment in a properly designed and operated purification plant. Color and odor are related to consumer acceptance of the water rather than to its safety. At 15 units of color or a threshold odor of 3, the water becomes objectionable to many people, and often they turn to alternate supplies of water for drinking that are less safe.

Presently water from many protected watersheds which are subjected to chlorination treatment only can meet the present Standard of 5 JTU (Jackson Turbidity Units) most of the time. However, this present concession to these unfiltered surface water supplies undoubtedly will be subjected to close scrutiny in any future updating of the Standards for several reasons. Mention has already been made of the fact that the pressures of civilization are making a myth of the "protected watershed," and it is just a question of time until it will be universally required that all surface water supplies be filtered as well as chlorinated.

With filtration by sand beds or other single-media filters, it is possible to produce consistently a water with turbidity of 1 JTU or less. Using modern coarse-to-fine filters in combination with application of a coagulant (alum or polymer) to the filter influent as a filter aid, and with continuous turbidity monitoring of the filter effluent, it is quite practical to maintain finished water turbidities of 0.2 JTU or less. With expert operation, turbidities of 0.05 to 0.1 JTU can be maintained. More than 100 plants in the U. S. are now using coarse-to-fine filters with continuous effluent turbidity monitoring and have routinely produced turbidities in the finished water to 0.2 JTU or even lower if desired. There are several advantages to production of low turbidity water. First of all, disinfection, particularly viral disinfection, of the low turbidity water is much more effective. Hudson has shown that the incidence of viral disease is much less in cities which produce high quality water (turbidity less than 0.2 JTU) than in cities employing less efficient filtration. It is current good practice to keep turbidities below 0.2 JTU, and eventually it is expected that this will be reflected in the Standards by a reduction in the upper limit for allowable turbidity.

Copper.   Copper in water is not a health hazard but may impart a taste at concentrations of 1 to 5 mg/l.

Carbon Chloroform Extract (CCE).   The material in water measured by the CCE test represents the organic matter not removed in water treatment. The test is useful as a screening procedure for undetected toxic materials. Toxic materials recovered include chlorinated insecticides, nitrobenzenes, aromatic ethers, and the like. The limiting concentration in the Standards for CCE is one which can be met by reasonably good treatment of the water without leaving excessive amounts of ill-defined chemicals in the finished water. At the limit stated, the organics are not considered to be a health hazard. When concentrations of CCE exceed 0.2 mg/l, the taste and odor of the water is poor.

Iron.   Iron in water is not likely to have any toxicologic significance, but is otherwise objectionable at concentrations above 0.3 mg/l. It imparts a yellowish color (when oxidized) and a bitter or astringent taste to water, and appreciably affects the taste of beverages. Iron produces a brownish stain on laundered goods, plumbing fixtures, and buildings in lawn sprinkler range. Iron is quite easily removed from water by proper treatment.

Manganese.   Neurologic effects of manganese which have been reported as a result of manganese dust inhalation have not been reported from oral ingestion in man. It appears that the limiting of manganese in water for esthetic and economic reasons will preclude any possible physiologic effects from excessive intake. Manganese imparts a brownish or purplish color to water and laundered goods (when oxidized), and stains plumbing fixtures. It also impairs the taste of coffee and tea. It is desirable for domestic use and imperative for some industrial purposes to remove all manganese from water. However, there are practical difficultes in reducing manganese to residual concentration less than 0.05 mg/l by treatment and in measuring such low concentrations.

Nitrate. Ingestion of water containing nitrates in excess of 50 mg/l (as NO_3) by infants may produce methemoglobinemia (blue baby conditions). Also breast-fed infants of mothers drinking such water may be poisoned. Cows drinking water containing nitrate may produce milk sufficiently high in nitrate to cause infant poisoning. If the nitrate concentration is sufficiently high, both adults and animals can be poisoned, but this has not been reported to be a problem from drinking water. There is no method for economically reducing the nitrate content of water. The usual public health practice is to warn the population of a community having a water supply high in nitrates of the dangers of infant feeding of such water, and to inform them of alternate sources of safe water for infant feeding.

Radium-226. Generally above-average levels of radium-226 intake occur only in unusual situations where the drinking water contains naturally occurring radium-226, as in the case of certain ground waters, or from pollution of a supply by industrial wastes containing radium. The limit of 3 $\mu\mu$c/l for drinking water was set with average intake from other sources in mind, and if the intake from sources other than water is unusually high, then lower levels in water may be necessarily imposed, using guidelines established by the Federal Radiation Council.

Strontium-90. To date, the principal source of human intake of strontium-90 is in food as a result of fallout from weapons testing. The limit of 10 $\mu\mu$c/l in water is substantially higher than the highest level of strontium-90 found in public water supplies to date.

Zinc. Zinc in water is not known to cause serious effects on health, however, concentrations in excess of 5 mg/l may render water unpalatable.

Nonfilterable Residue. The AWWA Goals state that water should be free of observable suspended particles or residue after settling, and this is the purpose of this limit. This is certainly a readily achievable goal which should without exception be met.

Microscopic and Nuisance Organisms. The organisms in this classification include larvae, bugs, crustaceans, algae, iron bacteria, sulfur bacteria, and slime growths. They are most apt to be found in unfiltered surface water supplies or as a result of aftergrowth in clearwells or water distribution systems. It is obvious that organisms visible to the naked eye should not be present in a quality water nor should the nuisance organisms which may impart such undesirable characteristics as taste, odor, or poor appearance even after water stands, is frozen, or is heated.

Aluminum.   At levels of aluminum exceeding 0.05 mg/l, precipitation may take place on standing, or in the distribution system. Turbidity and nonfilterable residue will be affected adversely.

Filterable Residue.   Low dissolved solids content is desirable, if precipitation is to be avoided in boilers or other heating units, to reduce sludge in freezing processes, and to reduce rings on utensils and precipitations on foods being cooked. The goal is for water to contain less than 200.0 mg/l of filterable residue.

Carbon Alcohol Extract (CAE).   This requirement is supplementary to the CCE requirement. The proportion of organic materials most commonly found in water is roughly in the proportion indicated by the goals stated for CCE and CAE. What the material in the CAE fraction is and whether it includes toxics may vary substantially from water to water.

Hardness.   In softening a hard water, there are several factors to consider in establishing the optimum residual hardness to be maintained including consumer costs involved in hard water use, treatment costs, and finished water stability (tendency to deposit or corrode). Experience has shown that water of 80 to 100 mg/l as $CaCO_3$ hardness is generally acceptable. Whatever hardness goal is selected, it should be maintained at a uniform level.

Alkalinity.   One of the AWWA water quality goals is aimed at a practical way to measure the stability of water in the distribution system by means of alkalinity tests. The change in alkalinity in water introduced into the distribution system and also after 12 hr at 130°F in a closed plastic bottle followed by filtration are the means employed in this measure of stability. The maintenance of calcium carbonate stability is the most effective method of preventing corrosion of iron water mains. Undersaturation, as shown by an increase in alkalinity under test conditions, indicates the possibility of iron pickup and development of red water. Oversaturation, as indicated by a drop of alkalinity, points out the probability of carbonate deposition in utensils, water heaters, household plumbing, and in water mains. Water conforming to the goal of not more than ± 1 mg/l change in alkalinity should minimize the problems of corrosion and deposition.

Coupon Tests.   Coupon insertion in pipelines is now recognized as a valuable method for evaluating corrosion rates under actual service conditions. Coupons directly measure simultaneously all of the factors (totalling more than 15) involved in corrosion including the physical factors of velocity and turbulence. Presently the amount of data on tolerable corrosion rates of the coupons as related to deterioration of pipelines is limited, but the figures given in Table 1-4 are based on the best available information. As the coupon tests become more widely used in the water works field, the acceptable limits can be better defined. Bean[7] has described one design for coupon test equipment for water mains.

Trace Elements and Compounds. Table 1-5 gives information on trace elements and compounds for which no limit was published in the 1962 Standards, but for which limits have been unofficially adopted by the Division of Water Hygiene in the intervening period. Also included in Table 1-5 are some elements for which a standard has not yet been adopted, but for which one is being considered by the committee revising the 1962 standards. Antimony is toxic but less so than arsenic. Beryllium in some of its salts is poisonous in occupational exposure. Bismuth is a heavy metal in the arsenic family and should be avoided in water supplies. Boron ingestion in large amounts can affect the central nervous system, and protracted ingestion may cause borism, as discussed earlier in this chapter. Molybdenum's acute or chronic effects are not well-known, but excessive intake may be toxic. Mercury ingestion for a protracted period or in large amounts can damage the brain or central nervous system. Nickel may cause dermatitis in sensitive people, and doses in the 30 to 75 mg (as the sulfate) range have produced toxic effects. Uranyl ion may cause damage to kidneys. The pesticides listed in Table 1-4 have severe, adverse health effects when ingested in large amounts. Small amounts accumulate, and long range effects are generally unknown. The organic phosphorus and carbamate pesticides are severe poisons affecting the central nervous system; ingestion of small amounts over time can harm the central nervous system. The herbicides listed in Table 1-4 are a group having toxic properties of a generally lower order than the pesticides listed. However, they should not be used without great care and should not be found in drinking water.

Sodium. In normal man, biological regulatory mechanisms operate in such a fashion that sodium taken into the body, in excess of those minute quantities actually utilized by properly functioning body organs, is excreted, primarily in the urine. These mechanisms adjust the output rapidly and precisely to conform with intake. There is no tendency to accumulate sodium ion in the body when salt intake is excessive.

For reasons which are not completely clear, the mechanisms which are responsible for maintaining a constant sodium content in the body despite wide variation in sodium intake fail to function properly in several very common diseases of the heart, kidney, and liver. Some patients with such diseases are unable to rid themselves of sodium and, consequently, large amounts of sodium-containing fluid (extracellular fluid) accumulate, a condition known as edema. It appears that the fluid retention largely may be prevented, or, if already present, dissipated by restriction of the amount of sodium which the person takes into his body. Another smaller group at high risk are the pregnant women who are suffering from toxemia of pregnancy. These patients have marked edema and store water and sodium in great excess. Depending on the type of disorder and the severity of the tendency to retain fluid, diets containing from approximately 250 mg to 2.5 g of sodium daily are prescribed.

One of the major dietary sources of sodium is water used for drinking and culinary purposes. If it is assumed that there is an average daily consumption of 2.5 liters of water in food and drink, it follows that when the water contains as much as 200 mg/l, the patient will receive 500 mg of sodium from water alone. While this sodium intake is not harmful to normal individuals, it must be considered for patients on a sodium restricted diet. One essential precaution to be taken in every instance is to ascertain whether or not the patient has installed in his home a zeolite water softener. The ordinary household softening unit is of the base-exchange type which softens the water by replacing the calcium and magnesium by sodium in the ratio of two atoms of sodium for each atom of calcium or magnesium. Home water softeners are seldom suitable for low sodium diets because the sodium content of the water is substantially increased. To illustrate this point, assume that a water has an initial sodium content of 387 mg/l and a total hardness of 1250 mg/l as $CaCO_3$. If this water were softened by the zeolite process, each liter of the water would then contain 960 mg of sodium.

Waters similar in character to the above, which are used for drinking and culinary purposes, would vitiate the results of sodium restriction in solid food intake and could be seriously detrimental to certain patients with cardiac, renal, or hepatic disease. It is essential to consider water used as a beverage and in the preparation of foods as a source of sodium in the establishment of a restricted sodium regimen in the treatment of the disorders indicated. Each case should be considered individually. Once the sodium content and fluid intake have been prescribed, the water supply available to the patient should receive careful consideration as a source of sodium independent of that from solid foods. In some instances, distilled, deionized, or low sodium surface waters may have to be substituted for drinking waters with a high sodium content in the dietary therapautic regimen for these patients.

Virus. Water meeting the bacteriological quality standards is believed to be safe from a viral standpoint. Reference has been made earlier in this chapter to the observations by Hudson pointing out the lower incidence of viral disease in cities where water treatment produced a superior product rather than a tolerable water. The specific relationship between finished water turbidities of less than 0.2 JTU and less incidence of viral disease was mentioned together with the probability that this is attributable to the better viral disinfection accomplished by chlorination in the low turbidity water. Because virus are so much smaller than bacteria, colloidal turbidity particles appear to offer more protection against contact with chlorine to virus than to bacteria, which is a possible explanation of the greater importance of turbidity removal to virus disinfection as compared to bacterial disinfection.

Since there is presently no method for positively detecting virus in water or measuring the numbers present, water works operators must rely on treatment procedures which are known to be efficient in virus removal. The principal tool is chlorination. Maximum efficiency of chlorination is dependent upon several factors including low turbidity, low pH, high free chlorine concentration, proper mixing, long contact time, and low concentrations of substances which exert a chlorine demand or other interference with the chlorination process. The importance of proper application and mixing of the chlorine and water being treated cannot be emphasized too strongly. It is recommended that the reader utilize the references at the end of this chapter to the work of Collins, Selleck, and White in this regard as well as reviewing material presented later in this text.

INDUSTRIAL WATER QUALITY REQUIREMENTS

General. The quality requirements for industrial water are generally consistent with general public demands for clear, attractive water of moderate mineral content, free from iron and manganese, but certain requirements may exceed those of public water supplies. Usually this means that the industry concerned must provide additional treatment at its plant site. However, if the industrial needs are known and considered in the municipal water treatment plant, they sometimes can be met by additional processing there and without the need for supplemental treatment by the industry.

Air Conditioning. Water for air conditioning purposes should be as cold as possible, low in hydrogen sulfide, iron, and manganese.

Baking. The additional water quality requirements for baking beyond the Standards are for total iron and manganese less than 0.2 mg/l, hydrogen sulfide less than 0.2 mg/l, and no taste or odor.

Boiler Feed Water. General boiler feed water requirements which exceed those for public water supplies are given in Table 1-6.

TABLE 1-6 GENERAL QUALITY TOLERANCES FOR BOILER FEED WATER

Boiler Pressure psi	Color	Dis-solved Oxygen ml/l	Hard-ness mg/1 as CaCO₃	pH	TDS mg/l	Al₂O₃ mg/l	SiO₂ mg/l	CO₃ mg/l	HCO₃ mg/l	OH mg/l
0-150	80	2.0	75	8.0+	3000–1000	5.0	40	200	50	50
150–250	40	0.2	40	8.5+	2500–500	0.5	20	100	30	40
250 and up	5	0.0	8	9.0+	1500–100	0.05	5	40	5	30

Brewing. For brewing light beer, special water quality requirements include, iron less than 0.1 mg/l, manganese less than 0.1 mg/l, TDS below 500, alkalinity under 75 mg/l as CaCO₃, hydrogen sulfide less than 0.2 mg/l, no taste, odor, no chlorine, sodium chloride less than 275 mg/l, and pH 6.5 to 7.0.

Canning. Water quality requirements for canning are hardness 25 to 75 mg/l as CaCO₃ (for legumes), iron under 2.0 mg/l, manganese less than 2.0 mg/l, hydrogen sulfide less than 1.0, and no taste or odor.

Carbonated Beverages. Recommended water quality includes turbidity less than 2.0 JTU, organic color plus oxygen consumed less than 10 mg/l, hardness under 200 mg/l as CaCO₃, iron less than 1.0 mg/l, manganese less than 0.2 mg/l, TDS less than 850 mg/l, alkalinity 50 mg/l as CaCO₃, no taste and odor, and hydrogen sulfide less than 0.2 mg/l.

Cooling Water. The total hardness of cooling water should be 50 mg/l as CaCO₃ or less, hydrogen sulfide less than 5 mg/l, and iron and manganese each less than 0.5 mg/l. The water should not be corrosive or slime forming.

Food Processing. Water from public water supplies meeting the Standards is generally satisfactory for food processing provided iron and manganese are each less than 0.2 mg/l and the water is free of objectionable taste and odor.

Ice Making. For ice making, water should have the following special characteristics: turbidity less than 5.0 JTU, color under 5, alkalinity as CaCO₃ 30 to 50, TDS less than 300 mg/l, total iron and manganese less than 0.2 mg/l, and silica less than 10 mg/l. Calcium bicarbonate is especially troublesome and magnesium bicarbonate tends to produce a greenish color in the ice. Free carbon dioxide is beneficial in preventing cracking of the ice. Calcium, magnesium, and sodium sulfates and chlorides should each be less than 300 mg/l in order to prevent white butts in the ice cakes.

Laundering. Public water supplies conforming to the Standards are satisfactory for laundry use, provided iron and manganese are each less than 0.2 mg/l. Hardness less than 50 mg/l as $CaCO_3$ is preferred.

Plastic Manufacture. To make clear, uncolored plastics, water is required to have turbidity and color each less than 2.0, iron and manganese each less than 0.02 mg/l, and TDS less than 200 mg/l.

Paper Making. To make the highest grades of light paper, the following water quality requirements should be met; turbidity and color each under 5, total hardness as $CaCO_3$ less than 50, TDS under 200, total iron and manganese under 0.1 mg/l, and uniformity of composition and temperature are desirable.

Rayon (Viscose) For pulp production, turbidity and color each should be less than 5, hardness less than 8 mg/l as $CaCO_3$ (hard water coats fibers), iron 0.05 mg/l, manganese not over 0.03 mg/l, TDS less than 100, total alkalinity under 50 and hydroxide alkalinity under 8 mg/l as $CaCO_3$, copper less than 5 mg/l, SiO_2 under 25 mg/l, and Al_2O_3 not over 8 mg/l. For rayon manufacturing, no iron or manganese can be tolerated, turbidity must be less than 0.3 JTU, and the pH must be in the range of 7.8 to 8.3.

Tanning. For use in tanneries, water should have a total hardness less than 135, total alkalinity under 135, and hydroxide alkalinity less than 8 mg/1 as $CaCO_3$. Iron and manganese each should be under 0.2 mg/l, and the pH = 8.0.

Textiles. General textile manufacture requires water with a total hardness less than 20 mg/l as $CaCO_3$ and iron and manganese each less than 0.2 mg/l. Residual alumina must be less than 0.5 mg/l in order to prevent uneven dying, as alum is a mordant. The chemical composition of the water should not vary. In production of cotton bandage, the color must be under 5 and calcium, magnesium, soluble organic matter, and suspended matter may be objectionable; and the iron and manganese content should be zero.

REFERENCES

1. Allen, H. E., and Bacon, C. W., "Rapid Determination Of Filterable Residue In Natural Waters," *Journal American Water Works Association,* p. 355 (1969).
2. American Water Works Association, "Standard Methods for the Examination of Water and Wastewater," 13th Edition, APHA, AWWA, and WPCF, Publ. by APHA, New York, 1970.
3. American Water Works Association, Task Group Report, "Effects of Synthetic Detergents on Water Supplies," *Journal American Water Works Association,* 1355 (1957).
4. American Water Works Association, *Water Quality and Treatment,* 3rd ed., McGraw Hill Book Co., New York, 1971.
5. American Water Works Association, ASCE, CSSE, "Water Treatment Plant Design," *Journal American Water Works Association,* (1969).
6. American Water Works Association, "Quality Goals for Public Water," Statement of Policy, *Journal American Water Works Association,* p. 1317 (1968).
7. Bean, E. L., "Progress Report on Water Quality Criteria," *Journal American Water Works Association,* p. 1313 (1962).
8. Bellack, E., "Arsenic Removal from Potable Water," *Journal American Water Works Association,* p. 454 (1971).
9. Bills, *et al.*, "Sodium and Potassium in Foods and Waters," *J. Amer. Dietetic Assoc.*, p. 304 (1949).
10. Braidech, M. M., and Emery, F. H., "The Spectrographic Determination of Minor Chemical Constituents in Various Water Supplies in the U. S.," *Journal American Water Works Association,* p. 557 (1935).
11. Bruvold, W. H. and Ongerth H. J., "Taste Quality of Mineralized Water," *Journal American Water Works Association,* p. 171 (1969).
12. Bruvold, W. H., Ongerth, H. J., and Dillehay, R. F., "Consumer Assessment of Mineral Taste in Domestic Water," *Journal American Water Works Association,* p. 575 (1969).
13. Calif. Dept. of Public Health, Bur. of San. Eng'g., "Wastewater Chlorination for Public Health Protection," Proc. 5th Annual San. Eng'g. Symposium, May, 1970.
14. Cangelose, J. T., "Acute Cadmium Metal Poisoning," *U. S. Navy Med. Bull.*, p. 39 and 408 (1941).
15. Clarke, N. A., *et al.*, "Virus in Water," AWWA Committee Report, *Journal American Water Works Association,* p. 491 (1969).
16. Clarke, N. A., and Weeks, J. D., "Analytical Procedures for Measuring Chemicals in the PHS Drinking Water Standards, 1962," *Journal American Water Works Association,* p. 660 (1970).
17. Cohen, M. J., Kamphake, L. J., Harris, E. K., and Woodward, R. L., "Taste Threshold Concentrations of Metal in Drinking Water," *Journal American Water Works Association,* p. 660 (1960).
18. Collins, H. F., Selleck, R. E., and White, G. C., "Problems in Adequate Sewage Disinfection," ASCE Natl. Specialty Conference on Disinfection, Univ. of Mass., Amherst, Mass.

19. Culp, R. L., and Stoltenberg, H. A., "Fluoride Reduction at LaCrosse, Kansas," *Journal American Water Works Association*, p. 423 (1958).
20. Culp, R. L. "Disease Due to 'Nonpathogenic' Bacteria," *Journal American Water Works Association*, p. 157 (1969).
21. Culp, R. L., "Possible Effects of Drinking High-Sodium Waters On Patients Under Restricted Sodium Therapy," Kansas State Board of Health Bulletin, 1954.
22. Davids, H. W. and Lieber, M., "Underground Water Contamination by Chromium Wastes," *Water & Sewage Wks.*, p. 528 (1951).
23. Derby, R. L., "Methods of Testing and Significance of Boron in Water," *Journal American Water Works Association*, p. 1449 (1936).
24. Eaton, F. M., "Boron in Soils and Irrigation Waters and Its Effects on Plants," *Tech. Bull. 448*, USDA, Wash., D.C., 1935.
25. Eichelberger, J. W. and Litchenberg, J. J., "Carbon Adsorption for Recovery of Organic Pesticides," *Journal American Water Works Association*, p. 25 (1971).
26. Ettinger, M. B., "A Proposed Toxicological Screening Procedure for Use in the Water Works," *Journal American Water Works Association*, p. 689 (1960).
27. FWPCA, "Enzyme Detergents in Water Supplies," *Journal American Water Works Association*, p. 640 (1969).
28. Goldblatt, E. L., Van Denburgh, A. S., and Marsland, "The Unusual and Widespread Occurrence of Arsenic in Well Waters of Lane County, Oregon," Lane County, Oregon Health Dept., 1963.
29. Goudey. R. F., "The U. S. Public Health Service Drinking Water Standards — From the Operator's Viewpoint," *Journal American Water Works Association*, p. 1416 (1943).
30. Goudey, R. F., "Solving Boron Problems in Los Angeles Water Supply," *Western Construction News*, p. 275 (1936).
31. Griffin, A. E., "Significance and Removal of Manganese in Water Supplies," *Journal American Water Works Association*, 1326 (1960).
32. Hoak, R. D., "Recovery and Identification of Organics in Water," *Intern. J. Air and Water Pollution*, p. 521 (1962).
33. Hudson, H. E. Jr., "High Quality Water Production and Viral Disease," *Journal American Water Works Association*, p. 1265 (1962).
34. Just, J. and Szniolis, A., "Germicidal Properties of Silver in Water," *Journal American Water Works Association*, p. 492 (1936).
35. King, P. H. et al, "Distribution of Pesticides in Surface Waters," *Journal American Water Works Association*, p. 483 (1969).
36. Lieber, M. and Welsch, W. F., "Contamination of Ground Water by Cadmium," *Journal American Water Works Association*, p. 51 (1954).
37. Lindstedt, D. K. and Kruger, Paul, "Vanadium Concentrations in Colorado River Basin Waters," *Journal American Water Works Association*, p. 85 (1969).
38. Lowe, H. and Freeman, G., "Boron Hydride (Borane) Intoxication in Man," *AMA Archives of Industrial Health*, p. 523 (1957).
39. Maier, F. J., "Defluoridation of Municipal Water Supplies," *Journal American Water Works Association*, p. 879 (1953).

40. McCabe, L. J., Symons, J. M., Lee, Roger D., and Robeck, G. G., "Survey of Community Water Supply Systems," *Journal American Water Works Association*, p. 670 (1970).
41. McKee, J. E., and Wolf, H. W., "Water Quality Criteria," Calif. State Water Quality Control Board, Publ. 3-A, 1963.
42. Metzler, D. F. and Stoltenberg, H. A., "The Public Health Significance of High Nitrate Waters As A Cause of Infant Cyanosis and Methods of Control," *Transactions of the Kansas Academy of Science*, 1950.
43. Moore, Edward W., "Physiological Effects of the Consumption of Saline Drinking Water," Progress Report to the Subcommittee on Water Supply, National Research Council, 1952.
44. Pluntze, James C., "Water Utilities — A Victim of Success," *Journal American Water Works Association*, p. 551 (1971).
45. Princi, F., *Metals-Cadmium in Oxford Loose Leaf Medicine*, Vol IV, Chap. XI, New York, 1949.
46. Russell, E. L., "Sodium Imbalance in Drinking Water," *Journal American Water Works Association*, p. 102 (1970).
47. Smith, O. M., "The Detection of Poisons in Public Water Supplies," *Water Works Engr.*, p. 1293 (1944).
48. Stanley, W. E., "Notes on Water Supply In North African Campaign," *J. New Eng. WW Assoc.*, Vol. LVIII, No. **4,** (1944).
49. Stokinger, H. E. and Woodward, R. L., "Toxicologic Methods for Establishing Drinking Water Standards," *Journal American Water Works Association*, p. 515 (1958).
50. Storrs, P. N., "Toxicological Significance of Arsenic in Drinking Water," Special Report U. S. Public Health Service, Feb. 2, 1962.
51. Straub, C. P. *et al.*," Strontium-90 in Surface Waters in the U. S.," *Journal American Water Works Association*, p. 756 (1960).
52. Taylor, Floyd B., "Trace Elements and Compounds in Waters," *Journal American Water Works Association*, p. 728 (1971).
53. USDA, "Salt Problems in Irrigated Soils," Bull. No. 190, 1958.
54. U. S. Public Health Service, "Mercury in Water Supplies," *Journal American Water Works Association*, p. 285 (1970).
55. U. S. Public Health Service, "U. S. Public Health Service Drinking Water Standards, P32," U. S. Govt. Printing Office, Wash., D.C., 1962.
56. U. S. Public Health Service, "Report of the Secretary's Commission on Pesticides and Their Relationship to Environmental Health," Dept. of HEW, U. S. Govt. Printing Office, Wash., D.C., Dec. 1969.
57. U. S. Public Health Service, "Manual for Protection of Public Water Supplies from Chemical Agents," United States Public Health Service Publ. No. 1071-J1, May, 1966.
58. Walton, Graham, "Emergency Treatment for Reducing Concentrations of Specific Chemicals in Public Water Supplies," Nat. Water Supply Research Lab., 1968.
59. Winton, E. F., Tardiff, R. G., and McCabe, L. J., "Nitrate in Drinking Water," *Journal American Water Works Association*, p. 95 (1971).
60. Lieber, M. *et al.*, "Cadmium and Hexavalent Chromium in Nassau County Ground Water," *Journal American Water Works Association*, p. 739 (1964).

61. American Water Works Association, "Rapid Tests for the Detection of Toxic Materials In Water," Willing Water Supplement No. 21, April, 1953.

62. Kamphake, L. J. and Kabler, Paul W., "Methods for the Detection and Identification of Chemical Warfare Agents in Aqueous Carriers," U. S. Public Health Service Publication No. 529, January, 1957.

63. Harmeson, R. H., Sollo, F. W. Jr., Larson, T. E., "The Nitrate Situation in Illinois," *Journal American Water Works Association*, p. 303 (1971).

64. Winton, E. F. and McCabe, L. J., "Studies Relating to Water Mineralization and Health," *Journal American Water Works Association*, p. 26 (1970).

65. Barnett, P. R., Skougstad, M. W., and Miller, J. M., "Chemical Characterization of a Public Water Supply," *Journal American Water Works Association*, p. 61 (1969).

66. Fishman, M. J., and Downs, S. C., "Methods for Analysis of Selected Metals in Water by Atomic Absorption," USGS Water Supply Paper 1540-C, Washington, D.C., 1966.

67. "Are you Drinking Biorefractories Too?," *Environmental Science and Technology*, **7**, p. 14 (Jan. 1973).

2 Shallow Depth Sedimentation

THEORY

A description of the settling paths of discrete particles in an ideal, rectangular basin is useful in understanding settling phenomena. In such an ideal basin, as defined by Camp,[1] the paths of all discrete particles will be straight lines and all particles with the same settling velocity will move in parallel paths. The settling pattern shown in Figure 2-1 would be the same for all longitudinal sections. All particles having settling velocities, v_s, greater than v_0 will fall through the entire depth, h_0, and be removed. The portion of particles with settling velocities v_s less than v_0 which will be removed is equal to the ratio of the velocities, v_s/v_0[1]. It can be seen from Figure 2-1 that particles with v_s less than v_0 could be completely removed if false bottoms or trays were inserted at intervals, h. Without such trays, a basin with a length much greater than L_0 would be required to capture these particles. It is apparent from Figure 2-1 that as the interval h is reduced, the size of basin required to remove a given percentage of the incoming settleable material decreases.

The above theory demonstrates that the removal of settleable material is a function of the overflow rate and basin depth and is independent of the detention period. This theory was proposed in 1904 by Hazen[2] when he pointed out that the proportion of sediment removed in a settling basin is primarily a function of the surface area of the basin and is independent of the detention time. He pointed out that doubling the surface area by inserting one horizontal tray would double the capacity of the basin. He felt that trays spaced at intervals as low as 1 in. would be very desirable if the problems of sludge removal could be resolved.

Camp[1] presented a design for a settling basin which would capitalize on these advantages. It had horizontal trays spaced at 6 in., the minimum distance he felt was permissible for mechanical sludge removal. The basin had a detention time of 10.8 min., a velocity of 9.3 fpm, and an overflow rate of 667 gpd/ft². Outlet orifices were used to distribute the flow over the width of the trays. In discussing this article, Eliassen[3] noted that although tray tanks had been used for many years in the chemical and metallurgical industries, they had been used in only a few water or sewage treatment systems. Camp[4] felt the lack of acceptance was due to reluctance of design engineers to depart from previous practice in size and shape of basins.

Theory indicates that the use of very shallow settling basins enables the detention time of the settling process to be reduced to only a few minutes in contrast to conventional settling basin designs which use 1–4 hr detention. Application of this theory offers tremendous potential for minimizing the size and cost of water treatment facilities. This chapter describes the techniques which are now available for successful application of this principle.

BASIC TUBE CLARIFIER SYSTEMS AVAILABLE

Essentially Horizontal

The two basic shallow depth settling systems now commercially available are illustrated in Figure 2-2.[5,6] These configurations are (1) essentially horizontal and (2) steeply inclined. The operation of the essentially horizontal tube settlers is coordinated with that of the filter following the tube settler. The tubes essentially fill with sludge before any significant amount of floc escapes. Solids leaving the tubes are captured by the filter. Each time the filter backwashes, the settler is completely drained. The tubes are inclined only slightly in the direction of normal flow (5 deg) to promote the drainage of sludge during the backwash cycle. The rapidly falling water surface scours the sludge deposits from the tubes and carries them to waste. The water drained from the tubes is replaced with the last portion of the filter backwash water, thus no additional water is lost due to the tube draining procedure. This tube configuration is applicable primarily to small plants (1 mgd or less in capacity) and is often used in package plant systems which are described later in this book.

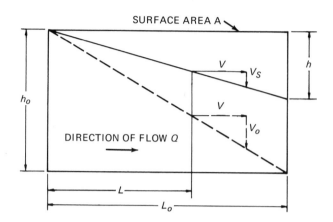

Fig. 2-1 Idealized settling paths of discrete particles in a horizontal flow tank.

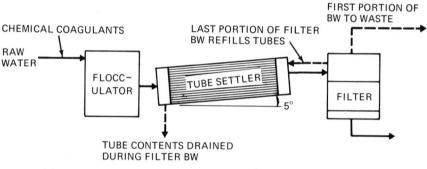

(a) ESSENTIALLY HORIZONTAL TUBE SETTLER

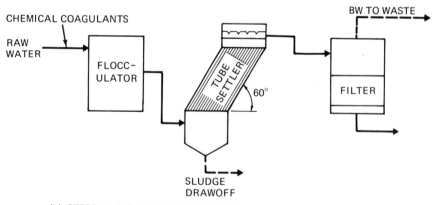

(b) STEEPLY INCLINED TUBE SETTLER

Fig. 2-2 Basic tube settler configurations shown schematically.

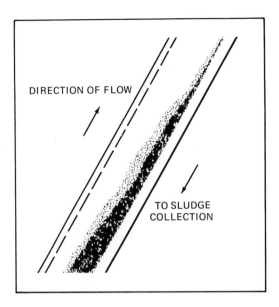

DIRECTION OF FLOW

TO SLUDGE
COLLECTION

Fig. 2-3   The flow of liquid and sludge quickly establish a countercurrent flow pattern in the steeply inclined tubes.

Steeply Inclined

Sediment in tubes inclined at angles in excess of 45 deg does not accumulate but moves down the tube to eventually exit the tubes into the plenum below. A flow pattern is established in which the settling solids are trapped in a downward flowing stream of concentrated solids (Figure 2-3). The continuous sludge removal achieved in these steeply inclined tubes eliminates the need for drainage or backflushing of the tubes for sludge removal. The advantage of shallow settling depth coupled with that of continuous sludge removal extends the range of application of this principle to installations with capacities of many millions of gallons per day.

Various manufacturers have developed alternate approaches for incorporating steeply inclined tubes into a modular form which is economical to build and can be easily supported and installed in a sedimentation basin. One modular construction is shown in Figure 2-4 (Neptune Microfloc) in which the material of construction is normally PVC and ABS plastic. Extruded ABS channels are installed at 60 deg inclination between thin sheets of PVC. By inclining the tube passageways rather than inclining the entire module, the rectangular module can be readily installed in either rectangular or circular basins. By alternating the direction of inclination of each row of the channels

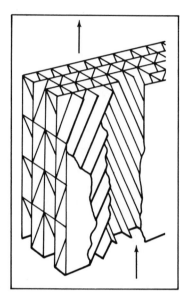

Fig. 2-4   A module of steeply inclined tubes. *(Courtesy Neptune Microfloc, Inc.)*

forming the tube passageways, the module becomes a self-supporting beam which needs support only at its ends. The rectangular, tubular passageways are 2 in. × 2 in. square in cross section and normally are 24 in. long.

Another manufacturer, Permutit, has adopted a chevron-shaped tube cross section as shown in Figures 2-5 and 2-6. The manufacturer feels that the V-groove in the bottom of the tube enhances the counterflow characteristics of the sludge. A third manufacturer, Graver Water Conditioning Company, utilizes rectangular channels similar to that shown in Figure 2-4 but with all of the channels inclined in the same direction.

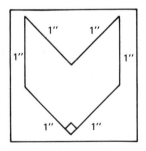

Fig. 2-5   Cross section of a chevron-shaped tube. *(Courtesy Permutit)*

Fig. 2-6 Modules of chevron-shaped tubes installed. (*Courtesy Permutit*)

While all of the above systems are used in configurations in which the influent is introduced beneath the tubes with the flow passing up through the tubes, one other system, the "Lamella" separator, has been recently introduced by the Parkson Corporation in which the influent enters at the top of the clarification basin and is then directed downward through a series of parallel plates (Figure 2-7). The sludge is collected at the bottom of the basin with the sludge-water flow being in the same direction rather than countercurrent as in the above systems. The clarified water is conveyed to the top of the clarifier by return tubes, as shown in Figure 2-7. The plates are typically 5 ft wide by 8 ft long, are spaced 1 1/2 in. apart, are inclined at 25-45 deg to the horizontal, and are usually constructed of PVC. Figure 2-8 depicts an installation as well as illustrating the discharge channel for the effluent return tubes.

At the time of this writing, there were not enough comparative data available to rank these various commercial configurations in terms of settling efficiency, maintenance costs, or cost effectiveness. The bulk of the experience has been gained with the first marketed system, that illustrated in Figure 2-4. Table 2-1 lists the major, operating installations of this type as of mid-1971. The following sections on design criteria are based largely on the experiences gathered from these installations.

INFLUENT

EFFLUENT

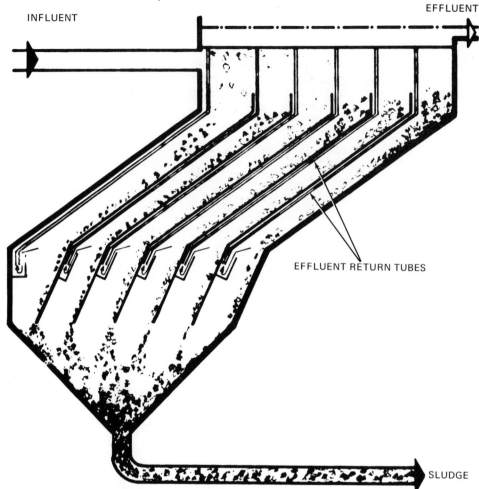

EFFLUENT RETURN TUBES

SLUDGE

Fig. 2-7   Lamella separator. *(Courtesy Parkson Corp.)*

GENERAL DESIGN CONSIDERATIONS

Steeply inclined tubes can be used in either upflow solids contact clarifiers or horizontal flow basins to improve performance and/or increase capacity of existing clarifiers. Of course, they can also be incorporated into the design of new facilities to reduce their size and cost. Capacities of existing basins can usually be increased by 50 to 150 percent with similar or improved effluent quality. The overflow rate at which tubes can be operated is dependent upon the design and type of clarification equipment, character of the water being treated, and the

Fig. 2-8 Lamella plates installed in concrete basin. *(Courtesy Parkson Corp.)*

desired effluent quality. The following sections describe the most important design and operational variables which affect tube installations in existing clarifiers.

Type of Clarifier

Horizontal Flow. The nature of the existing clarification equipment determines to some extent the allowable tube rate and the physical arrangement of modules in a basin. Ideal flow patterns are rarely experienced in practice in clarification basins. Velocities in rectangular, horizontal flow basins vary throughout the basin. Flow lines diverge at the inlet and converge at the outlet. The velocity gradient across the basin does not remain uniform due to basin drag, density currents, inlet turbulence, temperature currents, etc. In radial flow circular basins, the flow cannot be introduced to impart velocity components in the horizontal direction only. The use of a center feedwell imparts downward currents which cause turbulence and produces a general rolling motion of the contents in an outward and upward direction.

When installing tube modules in horizontal flow basins, it is best not

TABLE 2-1   PARTIAL LIST OF STEEPLY INCLINED TUBE CLARIFIER INSTALLATIONS (COURTESY NEPTUNE MICROFLOC INC.)

Name	Capacity mgd	Location	Date of Installation	New Plant	Existing Plant	Upflow Clarifier	Sed Basin	Engineer
Newport WTP	3.0	Newport, Oregon	1968		X	X		Cornell, Howland, Hayes & Merryfield
Bay Water Company	2.4	Pittsburgh, Calif.	1968		X	X		So. Calif. Water Company
Peace River WTP	1.4	Peace River, Alberta	1968		X		X	Associated Engineering Services Limited
Buffalo Pound WTP	28	Regina, Saskatchewan	1968–69		X	X		Associated Engineering Services Limited
Georgia-Pacific Corp.	45	Crossett, Arkansas	1968		X	X		Georgia-Pacific
Corvallis WTP	22	Corvallis, Oregon	1969		X		X	Cornell, Howland, Hayes & Merryfield
Flambeau Paper Company	8	Park Falls, Wisconsin	1969	X			X	S. J. Baisch
Emporia WTP	4	Emporia, Virginia	1969	X			X	Stuart Crawford Engrs.
Port St. John WTP	0.28	Port St. John, Florida	1969		X	X		General Development Utilities, Inc.
3 M Corporation	2	Middleway, West Virginia	1970		X		X	3 M Corporation
Portland WTP	1	Portland, Tennessee	1970		X	X		Howard, Nielsen, Lyne, Bates & O'Brien, Inc.
St. Albans WTP	2	St. Albans, West Va.	1970		X		X	Kelly, Gidley, Staub & Blair
Dresden WTP	1	Dresden, Ontario	1969		X	X		Dresden Utilities Commission
Stanton WTP	30	Stanton, Delaware	1970		X		X	Whitman, Requardt & Assoc.
Kirkwood WTP	2.5	Kirkwood, Missouri	1970		X		X	Horner & Shifrin
Ontario WTP	6.0	Ontario, Oregon	1971		X	X		Cornell, Howland, Hayes & Merryfield

Facility	Capacity (mgd)	Location	Year					Owner / Consultant
Caribou WTP	2.5	Caribou, Maine	1971			X	X	General Water Works
Venepal Pulp & Paper	4.5	Venepal, Venezuela	1970	X	X	X		Owner
Western Kraft Corp.	1.2	Albany, Oregon	1970	X	X	X		Owner
Petro-Tex Chemical Co.	1.0	Pasadena, Texas	1970			X	X	Owner
Loudon WTP	2.0	Loudon, Tennessee	1971		X		X	John C. Hayes & Assoc.
Waverly WTP	1.5	Waverly, Tennessee	1971		X		X	John C. Hayes & Assoc.
U. S. Navy	10.0	Subic Bay, Philippines	1971			X	X	Rogers Engineering
Boise Cascade	24.0	Internat'l. Falls, Minn.	1971		X	X		H. A. Simons, Ltd.
Petro-Tex Chemical Co.	8.6	Pasadena, Texas	1971			X	X	Owner
Ponca City WTP	6.5	Ponca City, Oklahoma	1971			X	X	City
Warsaw WTP	1.6	Rock Glen, New York	1971			X	X	Conable, Sampson, Kau Kurens, Huffcut & Gert
Clinton Corn Co.	19.4	Clinton, Iowa	1971	X	X	X	X	Owner
Tecumseh WTP	1.0	Tecumseh, Oklahoma	1971	X	X	X		Settle & Dougall Engineers
Saratoga Springs WTP	12.0	Saratoga Springs, N.Y.	1971			X	X	Rist-Frost Associates
Erie County Water Authority	24.0	Buffalo, New York	—			X	X	Malcolm Pirnie, Inc.
Pawhuska WTP	3.0	Pawhuska, Oklahoma	1971			X	X	C. H. Guernsey
Lawrenceburg WTP	3.0	Lawrenceburg, Kentucky	—			X	X	G. Reynolds Watkins

to locate them too near entrance areas where possible turbulence could reduce the effectiveness of the tubes as clarification devices. For example, in a horizontal basin, often as much as one third of the basin length at the inlet end may be left uncovered by the tubes so that it may be used as a sone for stilling of hydraulic currents. This is permissible in most basins since the required quantity of tubes to achieve a significant increase in capacity will cover only a portion of the basin. In radial flow basins, the required quantity of modules can be placed in a ring around the basin periphery, leaving an inter-ring of open area between the modules and the centerwell to dissipate inlet turbulence.

Upflow Clarifiers with Solids Contact.   The flow paths in solids contact basins of the upflow type are in a vertical direction through a layer or blanket of flocculated material which is held at a certain level and maintained at a certain concentration by the controlled removal of sludge. The clarification rate is governed by the settling velocity of this blanket. The purpose of maintaining this blanket is to entrap slowly settling, small particles which would otherwise escape the basin. When the flow is increased, the level of the blanket will rise. The efficiency of the tubes is dependent upon both the overflow rate and the concentration of incoming solids. The allowable loading rate on the tubes in this situation is dependent upon the average settling velocity of the blanket, the ability of the clarifier to concentrate solids, and the capacity of the sludge removal system to maintain an equilibrium solids concentration. If sludge is not withdrawn quickly enough or if the upward velocity exceeds the average settling velocity of the blanket, the unit can become solids critical with the result that the blanket will pass through the tubes with excessive carry-over of solids into the effluent.

In expanding the capacity of an upflow, solids contact clarifier, the ability to handle increased solids may be the limiting factor. The solids loading of the basin establishes the maximum capacity of the basin. The amount of increased capacity is often limited to 50 to 100 percent of the original capacity.

Basin Geometry

The shape of a basin determines how the tube modules can be most efficiently arranged to utilize the available space. The best arrangement may be determined strictly by basin geometry once the tube quantity is established. Of course, other factors must also be considered. For example, it may be desirable to locate the tubes as far as possible from areas of known turbulence or locate them so far to take advantage of an existing effluent launder system.

In circular basins, the tube modules are often placed in pie-shaped segments (see Figure 2-9). This approach is used where the entire clarification area is covered by tube modules or where the tube modules are placed in a ring around the basin outer wall. Where total coverage is required, the modules are supported by radial members which extend from an inner cone or ring to the outer wall of the basin. Where partial coverage is used and if the ring width does not exceed 10

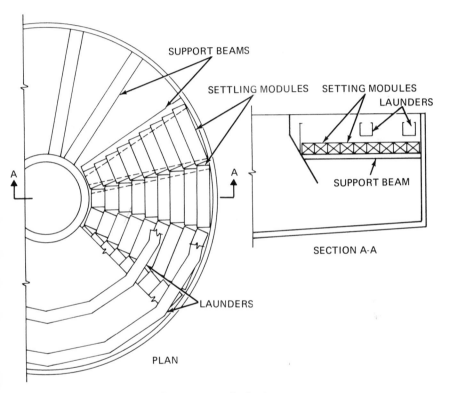

Fig. 2-9 Illustrative tube installation in circular basin.

to 12 ft, the support members may be cantilevered from the exterior walls. In any application where the entire area is not covered, a baffle wall must be installed at the inner perimeter of the modules to insure that all flow passes through the modules. Maximum width of a pie-shaped segment at the basin perimeter is limited by the maximum module length, which varies from manufacturer to manufacturer, but is usually on the order of 10–12 ft.

In basins which have radial effluent launders, it is often possible to suspend the modules from the launders, as illustrated in Figure 2-10. In

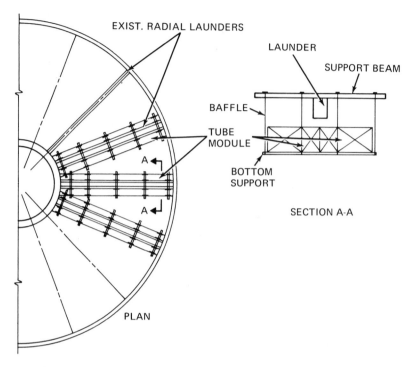

Fig. 2-10 Tube modules installed in conjunction with a radial effluent launder system.

rectangular basins, tubes are simply oriented with the long axis parallel to the sidewalls of the basin, with the support beams spanning the width of the basin (see Figure 2-11).

Tube Support Requirements

The tube support system must be able to support the weight of the tube modules when the basin is drained as well as making some allowance for the possibility of a workman standing on the modules. One manufacturer, Neptune Microfloc, recommends a surface loading of 7.5 lb/ft². The bearing surface width of a support member should be more than 1 in. to prevent possible shear failure of the module at the points of contact under extreme loading conditions. On the other hand, it should be as narrow as possible so as to block a minimum number of tube openings. The support members should be located a minimum of 6 in. (preferably 1 ft) in from the module end to develop the maximum support capability of the module.

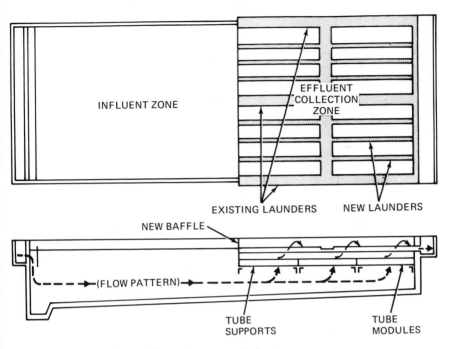

Fig. 2-11 Typical tube installation in rectangular basin.

Flocculation

When expanding a plant capacity with settling tubes, the existing flocculation facilities must be closely studied to insure that their capacity will not be overtaxed. The tubes are settling devices and cannot remove material which has not been flocculated to the point that is settleable.

Some older plants may have baffle flocculators which generally perform well at only one flow rate. The existing facilities, in this case, may have to be replaced or supplemented with more efficient mechanical units. In many cases, enough reserve capacity is built into existing mechanical flocculation units that the desired capacity increase can be achieved without modifying the flocculators. The exact detention time required depends on the energy input of the mixing device. A rough rule of thumb is that flocculation facilities may need upgrading if, at the increased plant flow, they provide less than 20 min detention at water temperatures below 45°F or 15 min with warmer temperatures. These detention times assume that the basins are well-designed so as to be free of short circuiting and that a polymer may be fed as a coagulant aid.

Sludge Removal Facilities

Some plants include settling facilities which were designed to be manually cleaned in cases where the raw water is of very low turbidity. These basins may be cleaned infrequently, perhaps only once per year, if the sediment load is very light. In other cases, they may be cleaned every 30 days. Although it is usually desirable to have continuous mechanical sludge collectors in basins equipped with tubes, tube modules have been successfully used in manually cleaned basins. The frequency of cleaning will obviously increase if the throughput through the basin is increased following the tube installation.

In general, the only adjustment required with mechanically cleaned water treatment basins will be to increase the frequency of withdrawal of the concentrated sludge from the basin. Most mechanical units have a substantial reserve capacity for handling larger quantities of sludge. In many plants, increasing the frequency of sludge withdrawal will merely involve the adjustment of a timer. In others, it may require complete modification or installation of supplementary sludge concentration and blowdown facilities.

Use of Coagulant Aids

As discussed elsewhere in this text, polymers may effectively increase the settling velocity of floc. The dosages required vary upon the nature of the specific water involved. However, to serve as a settling aid, dosages of 0.1 to 1.0 mg/l may be required. As with any type of clarifier, it is desirable to have the flexibility to feed polymers as a settling aid in basins equipped with tubes.

Tube Cleaning

In certain waters, floc has a tendency to adhere to the upper edges of the tube openings. This is generally an infrequent occurrence and of no serious consequence other than detracting from the appearance of the installation. In some cases, the floc buildup eventually bridges the tube openings and results in a blanket of solids on top of the tubes which may reach 3–10 in. in depth unless some remedial action is taken. One method of removing this accumulation is to drop the water level of the basin to beneath the top of the tubes occasionally. The floc particles are dislodged and fall to the bottom of the basin. In some cases, it may not be possible to remove the basin from service to drop the level. A very gentle water current directed across the top of the tubes by a fixed-jet header as shown in Figure 2-12 will remove floc accumulation. The header need be operated only infrequently, typically only a few minutes per day. Another cleaning technique involves

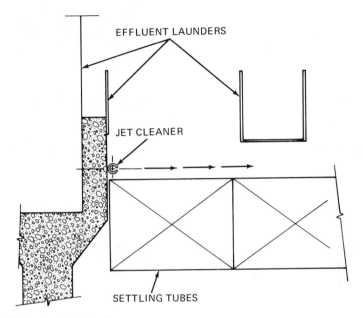

Fig. 2-12 Surface jet tube cleaner.

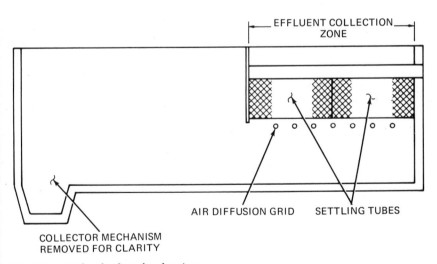

Fig. 2-13 Air header for tube cleaning.

the installation of a grid of diffused air headers beneath the tubes (Figure 2-13). To use this system, the influent is stopped and the air turned on and allowed to rise through the tubes, scrubbing any attached floc from the tubes. A quiescent period of 15–25 min follows before the basin is placed back in service.

RECOMMENDED DESIGN CRITERIA

Upflow Clarifiers — Loading Rates

In areas where cold water temperatures (less than 40°F) occur frequently, the following guidelines apply:

Overflow Rate, Based on Total Clarifier Area, gpm/ft²	Overflow Rate, Portion of Basin Covered by Tubes gpm/ft²	Probable Effluent Turbidity, JTU
1.5	2.0	1–3
1.5	3.0	1–5
1.5	4.0	3–7
2.0	2.0	1–5
2.0	3.0	3–7

In warm water areas (temperatures nearly always above 50°F), the following guidelines apply:

Overflow Rate, Based on Total Clariflier Area, gpm/ft²	Overflow Rate, Portion of Basin Covered by Tubes gpm/ft²	Probable Effluent Turbidity, JTU
2.0	2.0	1–3
2.0	3.0	1–5
2.0	4.0	3–7
2.5	2.5	3–7
2.5	3.0	5–10

Of course, these guidelines are based on the assumption that both the chemical coagulation and flocculation steps have been carried out properly. Also, the sludge removal equipment has been assumed adequate.

Horizontal Flow Basins — Loading Rates

As indicated in the tables below, the raw water turbidity has a direct influence on allowable tube overflow rates, as does the raw water temperature. In cold water areas where temperatures are frequently 40°F or less, the following guidelines apply:

Raw Water Turbidity = 0–100

Overflow Rated Based on Total Clarifier Area, gpm/ft²	Overflow Rate, Portion of Basin Covered by tubes, gpm/ft²	Probable Effluent Turbidity, JTU
2.0	2.5	1–5
2.0	3.0	3–7
3.0	4.0	5–10

Raw Water Turbidity = 100–1000

2.0	2.5	3–7
2.0	3.0	5–10

In warm water areas (temperatures nearly always above 50°F), the following guidelines apply:

Raw Water Turbidity = 0–100

Overflow Rate, Based on Total Clariflier Area, gpm/ft²	Overflow Rate, Portion of Basin Covered by Tubes gpm/ft²	Probable Effluent Turbidity, JTU
2.0	2.5	1–3
2.0	3.0	1–5
2.0	4.0	3–7
3.0	3.5	1–5
3.0	4.0	3–7

Raw Water Turbidity = 100–1000

2.0	2.5	1–5
2.0	3.0	3–7

Effluent turbidities above 5 JTU will often obscure the tube modules through 2 ft of water. This may not be aesthetically desirable to a casual observer but such turbidities are readily treated by a mixed-media filter. If the tube clarification application is not followed by mixed-media filters, the loading rates should be selected to provide effluent turbidities in the range of 1 to 3 JTU.

Location of Tube Modules Within the Basin

As discussed earlier, the tubes should be located such that they are not placed in a zone of unstable hydraulic conditions. Thus, they are frequently placed over 1/2–3/4 of the basin located nearest the effluent launders to permit the inlet 1/4–1/2 to dampen out hydraulic currents. The top of the tubes should be located 2 to 4 ft below the water surface. In general, the 2 ft minimum is used in shallow basins. The submergence of 4 ft would be considered only in clarifiers with a sidewater depth of 16 to 20 ft. In most basins, where sidewater depths rarely exceed 13 ft, a submergence of 2 to 3 ft is used. The collection launders should be placed on 10 to 12 ft centers over the entire area covered by tubes to insure uniform flow distribution.

Example Applications

Horizontal Basins. An existing water treatment plant with a rated capacity of 4 mgd has a horizontal, rectangular settling basin and rapid sand filters. The basin dimensions are 30 ft wide by 133 ft long. The surface overflow rate at design capacity is 1,000 gpd/ft². The average depth of the basin is 15 ft. The basin has a single overflow weir across the outlet end of the basin. The raw water is obtained from a river which has a normal maximum turbidity of 25–30 JTU. The water temperature rarely falls below 50°F. The settling basin is preceded by mechanical flocculation with 40 min of detention time at 4 mgd. Coagulant aids are fed during periods of high turbidity and low water temperatures to improve coagulation.

A capacity increase from 4 to 8 mgd is desired. At 8 mgd, the overflow rate increases to 2,000 gpd/ft² or 1.4 gpm/ft². The total basin loading of 1.4 gpm/ft² is below the values shown in the above guidelines for a horizontal flow basin under warm water conditions. However, at a higher basin loading of 2 gpm/ft² and a tube rate of 3 gpm/ft², the expected effluent turbidity is 1–5 JTU, which is compatible with mixed-media filters which, as discussed later in this text, can readily replace the existing sand filters. In light of the moderate raw water temperature and turbidity, a tube rate of 3 gpm/ft² and a basin loading rate of 1.4 gpm/ft² should give excellent results.

$$\text{Quantity of tubes required} = \frac{\text{Capacity, gpm}}{\text{Allowable tube rate, gpm/ft}^2}$$

$$= \frac{5600 \text{ gpm}}{3 \text{ gpm/ft}^2}$$

$$= 1870 \text{ ft}^2$$

The dimensions of the area to be covered by the tube modules is determined as follows:

Length of area covered by modules $= \dfrac{\text{Total area covered by tubes}}{\text{Width of basin}}$

$$= \frac{1870 \text{ ft}^2}{30 \text{ ft}}$$

$$= 62 \text{ ft}$$

The length of 62 ft would be rounded off to be a length readily compatible with the standard modules dimensions associated with the specific modules purchased.

The modules would be installed over an area extending back from the discharge end of the basin for a distance of 62 ft. A baffle wall would be installed at inner edge to force all flow through the modules. To provide uniform flow through the modules, 3 new effluent launders extending 62 ft back from the existing end wall launder would be required. The launders would be installed on 10 ft centers and the tubes would be submerged for a depth of 4 ft since the basin is deep. The appearance of the basin would be similar to that shown in Figure 2-11.

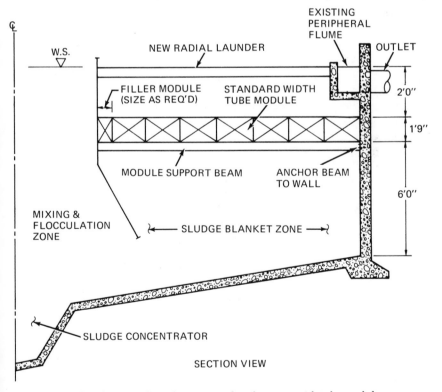

Fig. 2-14 Upflow basin with total coverage of surface area with tube modules.

Upflow Basins. A plant has two 42 ft square upflow clarifiers, each designed for a flow of 3000 gpm, with peripheral collection launders. The total surface area is 1760 ft². The influent centerwell reduces the available settling area by 200 ft². The peak overflow rate now reaches 1.92 gpm/ft², which is high enough that the clarifier does not perform well, especially when water temperatures drop.

It is desired to increase the plant capacity to 4000 gpm per settling basin. At this flow, the loading on the total basin settling area is 2.6 gpm/ft². The raw water turbidity is moderate, 30–70, and the water temperature seldom falls below 50°F.

The guidelines for upflow basins indicate that a maximum total basin loading of 2.5 gpm/ft² with a corresponding tube rate of 2.5–3 gpm/ft² can be used. Complete coverage of the settling area would provide a tube rate of 2.6 gpm/ft² and would also provide a simplified support problem when compared to only partial coverage. Thus, coverage of the settling area with 1560 ft² of tubes would be provided. Radial launders would be added to improve the flow distribution in the basin, as shown in Figure 2-14. As the sidewater depth is only 10 ft, the modules would be submerged 2 ft.

REFERENCES

1. Camp, T. R., "Sedimentation and the Design of Settling Tanks," *Transactions of American Society of Civil Engineers*, **111,** p. 895 (1946).
2. Hazen, A., "On Sedimentation," *Transactions of American Society of Civil Engineers*, **53,** p. 45 (1904).
3. Eliassen, R., Discussion of (1), *Transactions of American Society of Civil Engineers*, **111,** p. 947 (1946).
4. Camp, T. R. Discussion of (1), *Transaction of American Society of Civil Engineers*, **111,** p. 952 (1946).
5. Hansen, S.P., and Culp, G. L., "Applying Shallow Depth Sedimentation Theory," *Journal American Water Works Association*, **59,** p. 1134 (1967).
6. Culp, G. L., Hansen, S. P., and Richardson, G. H., "High Rate Sedimentation in Water Treatment Works," *Journal American Water Works Association*, **60,** p. 681 (1968).
7. Conley, W. R., and Slechta, A. F., "Recent Experiences in Plant Scale Application of the Settling Tube Concept," Presented at the Water Pollution Control Federation Conference, Boston, Mass., October, 1970.
8. Culp, G. L., Hsiung, K. Y., and Conley, W. R., "Tube Clarification Process, Operating Experiences," *Journal of the Sanitary Engineering Division*, ASCE, SA5, p. 829 (October 1969).

3 Filtration

DEFINITION

The term filtration has different meaning or connotations to various people. Outside the water works profession, even in other technical disciplines, filtration is commonly thought of as a mechanical straining process. It may also have this same basic meaning in water works practice as applied to the passage of water through a very thin layer of porous material deposited by flow on a support septum. This type of filter has a few rather specialized applications to water treatment as described later. However, most frequently in water works parlance, filtration refers to the use of a relatively deep (1½ to 3 ft) granular bed to remove impurities from water. This general type of filter has a wide range of applications. In contrast to mechanical strainers which remove only part of the coarse suspended solids, the newest types of filters used in water purification remove all suspended solids, including virtually all colloidal particles. Over the years the meaning of the term filtration as used by the water works industry has changed as improved filters have been developed and as the nature of the physical and chemical processes involved in filtration have become better understood. In an effort to distinguish recent new and improved filters from older conventional types, the term "separation bed" has sometimes been applied to mixed media filters which incorporate: (1)

51

coarse-to-fine in-depth filtration, (2) the application to the filter influent of a polymer, alum, or activated silica as a filter aid, (3) continuous monitoring of filter effluent turbidity, and (4) pilot filter control of coagulant dosage. However, separation beds are still referred to by many as "filters" and the important distinctions in functions and efficiency are often overlooked or not understood. In discussing filtration then, care must be exercised to make known specifically the details of the particular filter under consideration.

Water filtration can be further defined as a physical-chemical process for separating suspended and colloidal impurities from water by passage through a bed of granular material. Water fills the pores of the filter medium, and the impurities are absorbed on the surface of the grains or trapped in the openings. In nature, filtration is an active and important process in the purification of underground waters, and under controlled conditions in water purification plants, it is an indispensable unit process.

Filter Types

There are several ways to classify filters. They can be described according to the direction of flow through the bed, that is, downflow, upflow, biflow, radial flow, horizontal flow, fine-to-coarse, or coarse-to-fine. They may be classed according to the type of filter media used such as sand, coal (or anthracite), coal-sand, multilayered, mixed-media, or diatomaceous earth. Filters are also classed by flow rate. Slow sand filters operate at rates of 0.05 to 0.13 gpm/ft², rapid sand filters operate at rates of 1 to 2 gpm/ft², and high rate filters operate at rates of 3–15 gpm/ft². Another flow characteristic of filters is pressure or gravity flow. Gravity filter units are usually built with open top and constructed of concrete or steel while pressure filters are ordinarily fabricated from steel in the form of a cylindrical tank. Available head for gravity flow usually is limited to about 8 to 12 ft, while it may be as high as 150 psi for pressure filters. Because pressure filters have a closed top, it is not possible to routinely maintain a visual condition of the filter media. Further, it is possible to violently disturb the media in a pressure filter by sudden changes in pressure. These two factors have in the past tended to limit municipal applications of pressure filters to treatment of relatively unpolluted waters such as the removal of hardness, iron, or manganese from well waters of good bacterial quality. The susceptibility to bed upset and the inability to see the surface of the media in pressure filters have been compensated for to some extent at least by the use of quick-opening manholes and particularly by the recent development and application of recording turbidimeters for continuous monitoring of the filter effluent turbidity. The introduction of a

3 in. layer of coarse (1 mm) high density (specific gravity 4.2) garnet or ilmenite between the fine media and the gravel supporting bed has virtually eliminated the problem of gravel upsets and another of the concerns about the use of pressure filters for production of potable water.

Rapid or high rate filters operate at about 30 times the rate of slow filters, so they must be cleaned about 30 times as often. A common method for cleaning slow sand filters is to scrape a thin layer of media from the surface of the bed, wash it, and return it to the bed. Rapid filters are washed in place by reversing the flow through the media to expand and scour the media. This hydraulic cleaning process can be supplemented in various ways as necessary such as water jet agitation of the expanded media, mechanical stirring by rakes, or by injecting air into the bed before or during backwashing. With the scraping method used to clean slow sand filters, it was advantageous to collect as much of the foreign material as possible at the top surface of the bed. This favored the use of a fine-to-coarse filter medium. With modern hydraulic backwashing supplemented by water or air jet agitation, it is possible to thoroughly clean granular beds at all depths, making it feasible to tailor the graduation of the fine media (coarse-to-fine) for optimizing the filter cycle rather than the cleaning cycle of filter operation.

THE EVOLUTION OF FILTRATION

Slow Sand

In water works history there have been three basic types of filters which have dominated the field at different times. In the 1800s it was the slow sand filter. It incorporated sand with an effective size of about 0.2 mm. This very fine sand produced good quality water from applied water of low turbidity at rates on the order of 0.05 to 0.13 gpm/ft² of bed area. Because of the low surface rates, slow sand filters required large areas of land and were costly to install. They were also expensive to operate due to the laborious method of bed cleaning by surface scraping.

Rapid Sand

Beginning in the early 1900s under the stimulus of epidemic waterborne disease, the rapid sand filter came into general use and largely replaced the slow sand filter. Rapid sand filter media might vary in effective size from 0.35 to 1.0 mm, with a typical value being 0.5 mm. This type of fine media has demonstrated the ability to handle applied turbidities of 5 through 10 JTU at rates up to 2 gpm/ft^2. The introduction of prior chemical coagulation and hydraulic back-wash made possible the use of rapid rather than slow sand beds.

In backwashing single media sand beds, hydraulic grading of the sand grains occurs. The very finest sand accumulates at the top of the bed and the coarser particles lie below. More than 90 percent of the particulates removed are taken out in the top few inches of the bed. Once a suspended particle has penetrated this top layer of fine sand, its chances are greatly increased for passing through the entire bed, because the void spaces become larger and the opportunities for contact decrease as the particles travel downwards. This is a well-recognized limitation of the rapid sand filter.

Mixed-Media

However, it was not until the early 1940s under the stimulus of a critical wartime need to produce greatly improved water quality (basically much lower turbidity water) for processing radioactive materials at Hanford, Washington, that coarse-to-fine filtration had its beginnings under the leadership and direction of Raymond Pitman and Walter Conley. Development of the coarse-to-fine principle of filtration has taken place in two major steps. The first step was the development of the dual media filter which typically uses 24 in. of 1 mm anthracite coal above 6 in. of silica sand. Basically this provides a two-layer filter in which the coarse upper layer of coal acts as a roughing filter to reduce the load of particulates applied to the sand below. Because of the different specific gravities of the two materials (coal 1.4, sand 2.65), the coal of the proper size in relation to the sand remains on top of the sand during backwashing. With applied turbidities of less than about 15 Jackson Units, dual media filters can operate under steady state conditions at 4 to 5 gpm/ft^2 with the production of high quality water. Dual media filters can retain more material removed from the water than a sand filter, however they have a low resistance to turbidity breakthrough with changing flow rates. This serious shortcoming is due to the low total surface area of media particles, which is actually less than that for a conventional sand bed. Coal-sand beds in which there is considerable mixing of the two materials near the interface perform better and wash easier than coal-sand beds which are designed for more distinct layering (early versions).

Ideally, the size of the media particles and the pore space both should be uniformly graded from coarse-to-fine in the direction of flow through a filter bed. In a downflow filter, this would require almost an infinite number of materials, each having a different specific gravity. However a uniformly tapered void graduation and a uniformly tapered

MIXED MEDIA FILTER BEFORE AND AFTER INITIAL BACKWASHING

ORIGINAL PLACEMENT OF MATERIALS IN FILTER

COARSE COAL SP. GR. = 1.4

MEDIUM SAND SP. GR. = 2.65

FINE GARNET SP. GR. = 4.2

POSITION OF MATERIALS IN FILTER AFTER BACKWASHING

MIXED-MEDIA WITH PORE SPACE GRADED COURSE-TO-FINE IN DIRECTION OF FILTRATION (TOP TO BOTTOM)

16-MESH GARNET SUPPORTING SUPPORTING GRAVEL

Fig. 3-1 Mixed-media filter before and after initial backwashing.

average particle size can be obtained using only three properly graded materials: coal (specific gravity 1.4), sand (specific gravity 2.65), and garnet (specific gravity 4.2) approximately in the proportions of 60, 30, and 10 percent, respectively. The particle sizes range from 1.0 mm down to 0.12 mm from top to bottom of the bed. After backwashing, the three materials are mixed thoroughly throughout the depth of the bed (Figure 3-1). At each level in the bed are particles of coarse coal, medium sand, and fine garnet. The top of the bed is predominately coal, the middle is predominately sand, and the bottom is mostly garnet, but all three are present at all depths. In a properly designed mixed-media bed, the pore space uniformly decreases from top to bottom, the total number of particles at any level increases from top to

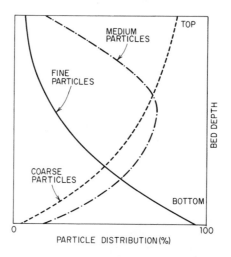

Fig. 3-2 Distribution of media in a properly designed mixed-media filter. *(Courtesy Neptune Microfloc, Inc.)*

bottom of the bed, and the *average* grain size decreases uniformly from top to bottom. Many persons have the misconception that a mixed-media bed is a three-layered bed. This is neither the case or the intent. A three-layered bed is not nearly as efficient. Dr. K. J. Ives has referred to mixed-media as "mixed up" media, which is perhaps the phrase which best describes the composition of the filter. The distribution of media in properly designed mixed-media bed is illustrated by Figure 3-2. Multimedia is another, less descriptive term, which is sometimes loosely applied either to mixed-media or multilayered filters.

Particles of turbidity in the influent to a mixed-media filter first pass through the large pores and encounter the coarse media, then reach the smaller pores resulting from the mixing of the finer media with the coarser media where more opportunities for contact occur. Materials are removed and stored throughout the full depth of the bed in contrast to the same functions in a sand bed which occur only in the top few inches of sand. The vast storage capacity of the mixed-media bed greatly increases the length of filter run before terminal head loss is reached. The total surface area of the grains in a mixed-media is much greater than for a sand or dual media bed which makes it much more resistant to breakthrough and more tolerant to surges in flow rates. This provides a great factor of safety in filter operation. Despite the greater total surface area of grains in mixed-media as compared to sand or dual media filters, the initial (clean filter) headloss in the two types are comparable. At 5 gpm throughput, the initial headloss in either a 0.50 mm sand or a mixed-media bed is about 1-½ ft in each.

Filter rates of 5 gpm/ft² are commonly used for design and operation of mixed-media beds as compared to 2 gpm/ft² for sand filters. At the same time, water quality is improved, which was the original purpose behind the development of the mixed-media bed. Along with the development and acceptance of the mixed-media filter has come a recognition that the rate of filtration is only one factor (and a relatively unimportant one) affecting filter effluent quality, and that chemical dosages for optimum filtration rather than maximum settling as well as other variables are much more important to production of good water.

In the modern concept of water treatment, coagulation and filtration are inseparable. They are actually each very closely related parts of the liquid-solids separation process. It is only because most water plants utilize sedimentation for a preliminary gross separation of settleable solids between coagulation and filtration that the crucial direct relationship of coagulation to optimum filtrability has been overlooked in the past.

Over the period of the last 10 years, the mixed-media filter has begun to gradually replace the rapid sand filter as the standard of the industry. The work started by Pitman and Conley at Hanford was aided in the early 1960s by the efforts of C. F. "Buck" Whetsler at Pasco, Washington; Ben Barnts, Ken Rinard, and Win Berkley at Eugene, Oregon; John McCool at Richland, Washington; R. L. Chandler and Cal Jackson at Johnson County, Kansas; Robert Lee at Medford, Oregon; and Archie Rice and others at CH2M/Hill in Corvallis, Oregon. In the late 1960s, the use of mixed-media filters spread to all parts of the world, and there are now hundreds of operating installations as discussed in more detail later. In view of the higher capacity, superior quality of water produced, and greater safety and reliability of the mixed-media filter as compared to other types, its current widespread use is not so surprising as the fact that more than 10 years elapsed after it was completely developed and fully demonstrated before it became so widely recognized and used. But, then, the acceptance of new ideas is excruciatingly slow in the water works field because of the public health considerations which are involved.

HOW WATER FILTERS ACT

Mechanisms

Several mechanisms are involved in particle removal by filtration. Part of the mechanisms are physical and others are chemical in nature. To fully explain the overall removal of impurities by filtration, the effects of both the physical and chemical actions occurring in a granular bed must be combined. Efficient filtration involves particle destabilization and particle transport similar to the mechanisms of coagulation. Good coagulants are also efficient filter aids. The processes of coagulation and filtration are inseparable and the interrelationships must be considered for best treatment results. One important advantage of filtration over coagulation in the removal of colloidal particles which are present in very dilute concentrations is the much greater number of opportunities for contact which are afforded by the granular bed as compared to the number afforded by stirring the water. The removal efficiency of a filter bed is independent of the applied concentration while the time of flocculation is concentration dependent.

The removal of suspended particles in a filter consists of at least two steps: (1) the transport of suspended particles to the solid-liquid surface of a grain of filter media or to another floc particle previously retained in the bed, followed by (2) the attachment and adsorption of particles to this surface. For filter runs of practical length, filter aid to coat filter grains must be added continuously (rather than pre-coated). When filter aids are used, the filtration process is so effective that conventional sand or other surface filters clog very rapidly. Effective filtration without excessive head loss can be accomplished best by use of a coarse-to-fine, in-depth filter.

There are two basic approaches to obtaining optimum filtrability of a water. One is to establish the primary dosage of coagulant for maximum filtrability rather than for production of the most rapid settling floc. A second approach is to add a dose of a second coagulant as a filter aid to the settled water as it enters the filter.

Adsorption of suspended particles to the surface of the filter grains is an important factor in filter performance. Physical factors affecting adsorption include both the nature of the filter and the suspension. Adsorption is a function of the filter grain size, floc particle size, and the adhesive characteristics and shearing strength of the floc. Chemical factors affecting adsorption include the chemical characteristics of the suspended particles, the aqueous suspension medium, and the filter medium. Two of the most important chemical characteristics are the electrochemical and van der Waals forces (molecular cohesive forces between particles).

For effective filtration, the objective of pretreatment should be to produce small, dense rather than large, fluffy floc so that the particles are small enough to penetrate the surface and into the bed. Removal of floc within a bed is accomplished primarily by contact of the floc particles with the surface of the grains or previously deposited floc, and adherence thereto. Contact is brought about principally by the convergence of stream lines at contractions in the pore channels between the grains. Of minor importance is the flocculation, sedimentation, and entrapment which occurs within the pores of the bed.

Filtration Efficiency

Filters are highly efficient in removing suspended and colloidal materials from water. Impurities affected by filtration included: turbidity, bacteria, algae, viruses, color, oxidized iron and manganese, radioactive particles, chemicals added in pretreatment, heavy metals and many other substances. Because filtration is both a physical and a chemical process, there are a large number of variables that influence filter efficiency. These variables exist both in the water applied to the filter, and in the filter itself. In view of the complexity of what is commonly considered to be a relatively simple process by plant designers and operators, it is surprising that the general level of filter performance is so high. Knowledge of the factors affecting filter efficiency increased quite rapidly during the 1960s. Use of this information in the design and control of filters will make possible remarkably better water quality with average or good filter operation, and will make it even more difficult to obtain unsatisfactory results with poor operation.

Filter efficiency is affected by the following properties of the applied water: temperature, filtrability, and the size, nature, concentration, and adhesive qualities of suspended and colloidal particles. Cold water is notably more difficult to filter than warm water, but usually there is no control over water temperature. Filtrability, which is related to the nature, size, and adhesive qualities of the suspended and colloidal impurities in the water, is the most important property. As will be discussed in more detail later, the only practical way to measure filtrability is by use of a pilot filter receiving raw water plus treatment chemicals and operating continuously in parallel with plant filters. By recording pilot filter effluent turbidities, appropriate adjustments can be made in chemical treatment to obtain optimum filtrability in the plant filters. The use of raw water in the pilot filter provides the necessary lead time to anticipate plant requirements. Maximum filtrability is much more important to production of a water of maximum clarity (minimum turbidity) than is maximum turbidity reduction prior to filtration.

Some properties of the filter bed which affect filtration efficiency are: the size and shape of the grains, the porosity of the bed (or the hydraulic radius of the pore space), the arrangement of grains, whether from fine-to-coarse or coarse-to-fine, the depth of the bed, and the headloss through the bed. In general, filter efficiency increases with smaller grain size, lower porosity, and greater bed depth. Coarse-to-fine filters contain much more storage space for materials removed from the water and permit the practical use of much finer materials in the bottom of the bed than can be tolerated at the top of a fine-to-coarse filter. Because of the greater total surface area of the grains, the smaller grain size and lower porosity, the coarse-to-fine filter is more efficient than fine-to-coarse filter. The much greater total grain surface area and the smaller grain size provided by mixed-media as compared to dual media account for the greater resistance to breakthrough possessed by mixed-media. Dual media which should be considered merely an intermediate step in the development of mixed-media, is less resistant to breakthrough than a rapid sand filter while mixed-media is more resistant than either. Hydraulic throughput rate also affects filter efficiency. Although, as previously pointed out, within the range of 2 to 8 gpm/ft^2, the rate is not nearly as significant as other variables are to effluent quality. In general, the lower the rate, the higher the efficiency. All other conditions being equal, a filter will produce a better effluent when operating at a rate of 1 gpm/ft^2 than when operating at 8 gpm/ft^2. However, it is also true that a given filter may operate entirely satisfactorily at 8 gpm/ft^2 on a properly prepared water, yet fail to produce a satisfactory effluent at 1 gpm/ft^2 when receiving an improperly pretreated water. With good filter design, the optimum throughput rate is a matter of economics rather than a question of safety.

The efficiency of filters in bacterial removal varies with the applied loading of bacteria, but with proper pretreatment should exceed 99 percent. Bacterial removal by filtration, however, should never be assumed to reach 100 percent. The water must be chlorinated for satisfactory disinfection. Coagulation, flocculation and filtration will remove more than 98 percent of polio virus at filtration rates of 2 to 6 gpm/ft^2, but complete removal is dependent upon proper chlorination.

The turbidity of the effluent from a properly operating filter should be less than 0.2 JTU. With proper pretreatment, filtered water should be essentially free of color, iron, and manganese. Large microorganisms, including algae, diatoms, and amoebic cysts are readily removed from properly pretreated water by filtration.

The perfection of very sensitive, accurate, and reliable turbidimeters for the continuous monitoring of filter effluent quality has revolutionized the degree of control that can be exercised over filter performance. The turbidity of filter effluent is instantly and continuously determined, reported, and recorded. The signal obtained from the instrument can be used to sound alarms, or, if necessary, shut down an improperly operating filter unit. This greatly increases the reliability with which filters of all types may be operated, and may broaden the range of safe application for pressure filters.

The Role of Filtration in Water Treatment

Today, it is generally acknowledged that all surface water supplies should be filtered. There are still a significant number of surface water supplies derived from uninhabited, "protected," watersheds which are not filtered. These unfortunate conditions result from applying economic expediency rather than the best technical judgment to these situations, and it is now only a matter of time before all public water supplies obtained from surface sources will be required to provide the essential safeguard of filtration.

Some well waters meet water quality standards and goals without treatment other than chlorination. On the other hand, many well waters require treatment for removal of iron, manganese, hardness, color, odor, turbidity, hydrogen sulfide, bacteria, virus, or other undesirable impurities. In these cases, filtration is ordinarily a part of the overall purification process.

In water treatment, the general practice is to reduce the turbidity of water to 10 JTU before application to rapid sand filters. However, mixed-media filters can and are handling applied turbidities up to 50 JTU on a continuous basis, and will tolerate occasional turbidity peaks of 200 JTU. This is a relatively new capability of filters which should not be overlooked in selecting plant flow sheets. Several public water supplies derived from surface sources having average turbidities of 20 to 50 JTU and peaks of 100 to 200 JTU are operating quite successfully without the use of settling basins ahead of the filters.

Provisions should be made in plant design to chlorinate both filter influent and effluent. Chlorine applied ahead of a filter is a great aid in maintaining clean filters. Chlorination of the finished water makes the most effective use of chlorine because of the reduced chlorine demand at this point in treatment. Provisions should also be made for application of a filter aid such as alum, activated silica, or a polymer ahead of the filters so that the water can be adjusted to optimum filtrability.

Where there is no full plant operating experience with a particular water to be treated, or where the records are not adequate to determine the best treatment processes to be used, pilot plant studies are a valuable aid to producing the most efficient and economical scheme of treatment. They are highly recommended under these conditions. Pilot studies can be used to determine the kind and extent of pretreatment which may be required ahead of filtration. Only a filter itself can reveal through its operation the characteristics of floc particles and filter media that bear on removal efficiency and length of filter runs. While backwash rates are well-standardized, it is good to check them through pilot plant operation on a particular water to be treated. In conducting pilot plant tests, the necessity to operate under all raw water conditions must not be overlooked. Unless the pilot plant is operated under the most severe conditions, the full scale plant may be underdesigned and ill-equipped to produce the desired water quality at all times, as is necessary.

The Value of Filter Formulas

Filtration is one of the most widely used and extensively investigated processes in the field of sanitary engineering. Yet there is not at this time a clear understanding nor an exact formulation of the mechanisms by which particle removal from water is accomplished by filtration. Not one of the theories of filtration can predict in advance the performance of a particular filter treating a given suspension. This can only be done by extensive laboratory, pilot plant, or full scale testing. In the past, all of the major advances in filtration have come about as a result of the efforts of men like Fuller, Baylis, and Conley who, although they were scientists, were principally practitioners of the art of filtration rather than theoreticians or mathematicians. Current theories of filtration are either too simple to be widely applicable, or too complex to be useful. They are too complex in the sense that they require an extensive amount of experimentation to determine the formula constants which are applicable to a given situation. The various theories and formulas disagree to such an extent that their general applicability is obviously questionable.

Many theories deal only with the physical aspects of filtration such as media, rate, and water temperature. Other theories consider only the chemical aspects such as surface characteristics of the suspended particles and the chemical characteristics of the aqueous phase.

There is no doubt that water filtration is affected by both physical and chemical parameters. Filtration is a combination of particle transport and attachment. Particle transport is probably primarily influenced by the size and density of the suspended particles, and to a lesser extent by filtration rate, sand size, and fluid temperature. Particle attachment is brought about by the forces of colloid chemistry such as diffuse layer interactions, but the chemical effects on filter performance are diverse, interrelated, quite complicated, and very significant.

Analyses of the effects of physical parameters can be compared only if the effects of all chemical factors are constant. Likewise, investigations of the effects of chemical parameters are valid only if the physical factors are constant. To date, all attempts have failed to develop an exact mathematical formulation, with theoretical constants, for the filtration process to hold for any and all conditions of operation. The tasks and values of the theory are to identify the variables, to provide the base for rational experimental or test procedures, and to develop rational methods for data analysis in order to get the answers needed for engineering design and plant operation. For example, the Fair-Hatch formulas for filter backwashing can be used to approximate the relative vertical positions of the various sizes, shapes, and densities of the coal, sand, and garnet grains contained in a mixed-media bed when expanded. The exact relative grain positions in the settled bed following backwash cannot be calculated because of hindered rising and hindered settling effects, but computations do give a good base for starting experimental work.

Walter Conley and his co-workers at Neptune Microfloc have, over the past 10 or 12 years, conducted filtering and backwashing tests in the laboratory and field on literally thousands of mixed-media bed designs in bringing the mixed-media filter to its present state of perfection. This work continues and will enable not only improvement in media composition and bed design, but also refinement of the formulas through determination of the formula constants under different conditions.

DESIGN OF FILTER SYSTEMS

Selection of Type, Size and Number of Units

In designing new plants, the selection of type is not influenced by existing facilities, and the gravity, downflow, mixed-media bed in a concrete box with hydraulic surface wash, gravel and garnet supporting media, and Wheeler, Leopold, or pipe lateral underdrains will usually be the filter of choice. In small installations, the use of steel tanks or pressure filters may be indicated. These conclusions are based on information presented previously plus additional data which follow.

In the expansion of existing plants, there ordinarily is no problem in adding new units as described in the preceding paragraph, but there may be a question of the necessity or advisability of converting or partially converting existing filter units to match the new units.

In medium and large plants, the minimum number of filter units is usually 4. In small plants it may be 2, or even 1 if adequate finished water storage is provided to supply service demands and backwash needs during filter shutdown for backwashing. The maximum size of a single filter unit is limited by the rate at which backwash water must be supplied, difficulties in providing uniform distribution of backwash water over large areas, reduction in filter plant capacity with 1 unit out of service for backwashing, and structural considerations. The largest filter units in service are about 2100 ft^2 in area. Units larger than about 1000 ft^2 are usually built with a central gullet so that each half of the filter can be backwashed separately.

To filter a given quantity of water, the capital cost of piping, valves, controls, and filter structures is usually less for a minimum number of large filter units as compared to a greater number of smaller units. In expanding existing plants, it may be better not to increase the size of filter units but rather to match the existing size in order to avoid the need to increase the capacity of wash water supply and disposal facilities.

Filter Layout

One popular arrangement is to place filters side by side in 2 rows on opposite sides of a central pipe gallery. One end of the rows of filters should be unobstructed to allow construction of additional units in future expansion. The best pipe gallery design includes a daylight entrance. This provides good lighting, ventilation, and drainage, and improves access for operation, maintenance, and repair.

Filters should be located as close as possible to the source of influent water, the backwash water supply, the filtered water storage reservoir, and the control room. On sites which are subject to flooding, the level of the bottom of the filter boxes should be located above the maximum flood level. This permits the discharge of filter backwash water during flood periods, and avoids possible contamination of the filtered water.

In the past, the overall depth of filters from water surface to underdrains was ordinarily 8 to 10 ft. The current trend is toward deeper filter boxes in order to reduce the possibility for air binding and to increase the available head for filtration. Filter effluent pipes and conduits must be water sealed to prevent the entry of air into the bottom of the filter when it is idle.

The use of common walls between filtered and unfiltered water should be avoided in order to eliminate the possibility of contamination of the finished water. For the same reason, it is better not to construct clearwell storage beneath the pipe gallery floor. Many states do not permit these types of construction.

Filter Structures

The general practice is to house and roof pipe galleries. In the past, this was also the practice for filters and filter operating galleries. This was done to protect the filter controls and the water in the filters against freezing. Current practice is to locate the filter controls in a central control room and to provide perforated air lines around the periphery of each filter to control ice formation. This eliminates the need for housing the filters and filter gallery. If visual observation of filter washing without exposing the operator to the weather is considered essential, then it becomes a question of the relative costs and convenience of closed circuit television *vs* housing the filters and filter gallery.

For gravity filter structures, either concrete or steel may be used. Concrete boxes may be either rectangular or circular. Steel tanks are usually circular. For small tanks and pressure filters, steel is usually preferred. For large filters, concrete is the choice.

There are a number of special structural considerations involved in reinforced concrete filter boxes which deserve the attention of designers. They must carry not only wind, dead and live loads, but hydrostatic loads as well. To be watertight, walls should be not less than 9 in. thick. Walls must often be designed to withstand water pressure from either of two directions, and to resist sliding and overturning. Slabs also may be subject to hydrostatic uplift as well as gravity loads. Shear and watertightness require special attention particularly at keyways at the base of walls. The deflection at the top of cantilever walls and sliding at the base are common sources of trouble in water bearing concrete structures. The concrete should be impervious to minimize the deleterious effects of freezing and thawing, and ample cover should be provided for steel reinforcement. The difference in expansion between water-cooled walls and floors and exposed slabs and roofs must be considered. The location and design of construction and expansion joints must be shown on the plans by the engineer and not left to the judgment of the field inspector or contractor.

Filter Underdrains

Filter underdrains have a two fold purpose. The most important is to provide uniform distribution of backwash water without disturbing or upsetting the filter media above. The other is to collect the filtered water uniformly over the area of the bed. Ideally, the filter underdrain system would also serve as a direct support for the fine media. There is one type of underdrain (porous plates) which can do this under certain special conditions as will be described later. However, the three most widely used underdrain systems (Wheeler, Leopold, and pipe lateral) all require an intermediate supporting bed of gravel which should be topped with a layer of coarse garnet or ilmenite to prevent upsets of the gravel layer. Figures 3-3 and 3-4 illustrates two of the underdrain systems mentioned above. All of the systems accomplish the uniform distribution of wash water by introducing a controlling loss of head, usually about 3 to 15 ft, in the orifices of the underdrain system. The orifice loss must exceed the sum of the minor (manifold and lateral) head losses in the underdrain to provide good backwash flow distribution.

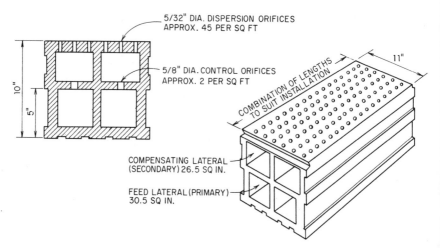

Fig. 3-3   Leopold filter bottom. *(Courtesy F.B. Leopold Co.)*

The Wheeler bottom can be either a pre-cast or poured-in-place false concrete bottom with a crawl space (22 in. high) between the structural and false bottoms. The crawl space and access manhole from the pipe gallery provide for very low headloss flow distribution and means for removal of the concrete form. The bottom consists of a series of conical depressions at 1 ft centers each way. At the bottom of each conical depression is a porcelain thimble with a 3/4 in. diameter orifice opening. Within each cone there are 14 porcelain spheres ranging in diameter from 1 3/8 to 3 in. and arranged so as to dissipate the velocity head from the orifice with minimum disturbance of the gravel above. Steel forms for the poured-in-place Wheeler bottom can be rented from either B-I-F, Providence, R. I., or Roberts Filter Co., Darby, Pa.

The Leopold bottom consists of vitrified clay blocks each about 2 ft long by 11 in. wide by 10 in. high. The bottom half of the block contains two feeder channels (about 4 in. $\times$ 4 in.) with two $5/8$ in. diameter orifices per ft^2 which feed water into the two dispersion laterals (about $3\frac{1}{4} \times 4$ in.) above with a headloss at a flow of 15 gpm/sf of about 1.45 ft. In the top of the block there are approximately forty-five $5/32$ in. diameter orifices per ft^2 for further distribution of the wash water and for dissipation of the velocity head from the lower orifices. These blocks and installation services are available from F. B. Leopold Co., Zelienople, Pa. or Warminster Fiberglass Co., Southampton, Pa.

The perforated pipe lateral system uses a main header with several pipe laterals on both sides. The laterals have perforations on the underside so that the velocity of the jets from the perforations during backwash is dissipated against the filter floor and in the surrounding gravel. Piping materials most commonly used are steel, transite, or PVC.

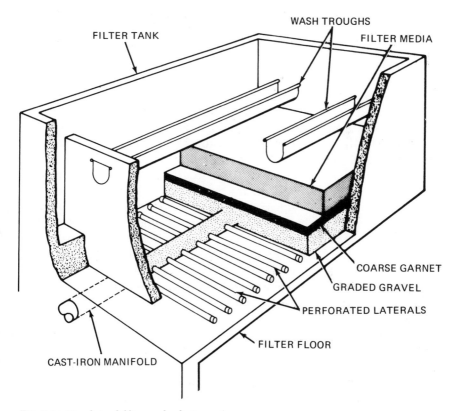

Fig. 3-4 Pipe lateral filter underdrain system.

Usually, orifice diameters are ¼–½ in. with spacings of 3 to 8 in. Fair and Geyer (1958) give the following guides to pipe lateral underdrain design:

1. Ratio of area of orifice to area of bed served, 0.15 to 0.001.
2. Ratio of area of lateral to area of orifices served, 2:1 to 4:1.
3. Ratio of area of main to area of laterals served, 1.5:1 to 3:1.
4. Diameter of orifices, ¼ to ¾ in.
5. Spacing of orifices, 3 to 12 in. on centers.
6. Spacing of laterals, about the same of spacing of orifices.

Several underdrain systems have been developed with the goal of eliminating the need for the gravel support bed. These systems employ strainer-like nozzles or porous plates. Plastic or steel nozzles of various shapes are available from several manufacturers. Among these are the Eimco Corporation, which manufactures a plastic, conical nozzle with a screen supported between two layers of plastic. Degremont Company makes plastic nozzles which have finely slotted sides. The Edward Johnson Company manufactures a stainless steel nozzle consisting of 4¾ in. diameter well screen pipe with a cap welded on one end. Walker process offers a nozzle made of thermoplastic resin having 0.25 mm aperature vertical slots. The nozzles are mounted in hollow, vitrified clay blocks laid to form the filter bottom. The 9 in. high hollow-core blocks are grouted in place and provide a system of ducts for filtered water flow as well as for backwash flow. The nozzles are designed with built-in orifices to meter the backwash flow evenly over the filter area. Care must be taken in selection of these nozzles to insure that the finest filter media will not pass through the nozzle openings. The nozzles are mounted either in a false bottom or in a pipe lateral. A disadvantage of the plastic nozzles is that they are fairly fragile and can be easily broken when being installed or when the medium is placed. An operating problem in use of the nozzles is plugging from the inside with material found in the backwash water or with rust or other particulate matter from the filter backwash piping. This plugging frequently occurs on the first application of backwash due to the failure to adequately clean the construction debris from the system. Some nozzles made from a combination of plastic and steel have cracked and the underdrain systems have failed because of differential contraction with temperature changes.

Porous plates made of aluminum oxide, such as manufactured by the Carborundum Company, are also available to eliminate the need for the gravel layer. The aluminum oxide plates are usually mounted on steel or concrete frames which support the plate on at least two edges, or by studs projecting from the filter floor fastened to the corners of the plates. Mastic or caulking compounds are used to seal the joints. The disadvantages of the aluminum oxide plates are that they are brittle and easily broken during installation; also, the sealing of joints between plates may be difficult and the plates can become plugged from either side. If plugged plates are not promptly cleaned, structural failure may occur due to decreasing differential pressures across the plates during backwash. They offer the advantages of good flow distribution and minimum filter height since the media can be placed directly on the plates. They may not be suitable for use in filtering water which deposits calcium carbonate due to plugging problems under this condition.

A decision to use a nozzle or porous plate underdrain system should be based on the personal satisfactory experience of the designer in similar applications. In general, currently available direct support systems have not yet proven to be as satisfactory as gravel support beds for fine media.

Filter Gravel

A graded gravel layer usually 14–18 in. deep is placed over the pipe lateral system to prevent the filter media from entering the lateral orifices and to aid in distribution of the backwash flow. A typical gravel design is shown in Table 3-1. A weakness of the gravel-pipe

TABLE 3-1   TYPICAL GRAVEL BED FOR PIPE UNDERDRAIN SYSTEM

Description		Number of Layer		
	Bottom	2	3	Top[b]
Depth of layer (in.)		3	3	4
Sq. mesh screen opening (in.)				
Passing	a	3/4	1/2	1/4
Retained	3/4	1/2	1/4	1/8

[a]Bottom layer should extend to a point 4 in. above the highest outlet of wash water.
[b]Plus coarse garnet.

lateral system has been the tendency for the gravel to eventually intermix with the filter media. These gravel "upsets" are caused by localized high velocity during backwash, introduction of air into the backwash system, or use of excessive backwash flow rates. The gravel layer can be stabilized by using 3 in. of garnet or ilmenite as the top layer of the gravel bed. This coarse, very heavy material will not fluidize during backwash and provides excellent stabilization for the gravel. It also prevents the fine garnet or ilmenite used in a mixed, trimedia filter from mixing with the gravel support bed. The remaining major disadvantage of the gravel-pipe lateral system is the vertical space required for the gravel bed. Gravel layers are also used with

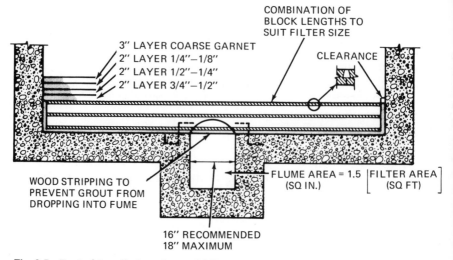

Fig. 3-5 Typical installation of Leopold filter bottom. *(Courtesy F.B. Leopold Co.)*

several of the commercially available underdrain systems such as the Leopold Bottoms (see Figure 3-5), and Wheeler Bottoms. Gravel depths and gradations vary for these underdrain systems. For example, Leopold recommends gradation listed in Table 3-2.

TABLE 3-2   GRAVEL SIZE AND LAYER THICKNESS FOR USE WITH CLAY TILE BOTTOMS

Gravel Layer	Layer Thickness (in.)	Size Limit (in.)
Bottom	2	$3/4 \times 1/2$
Second	2	$1/2 \times 1/4$
Top[a]	2	$1/4 \times 1/8$

[a]Plus coarse garnet.

The gravel to be used with Wheeler Bottoms is shown in Table 3-3.

TABLE 3-3   GRAVEL SIZE AND LAYER THICKNESS FOR USE WITH WHEELER BOTTOMS

Gravel Layer	Layer Thickness (in.)	Size Limit (in.)
Bottom	As required to cover the underdrain systems	$1 \times 1\frac{1}{4}$
Second	3	$5/8 \times 1$
Third	3	$3/8 \times 5/8$
Top[a]	3	$3/16 \times 3/8$

[a]Plus coarse garnet.

Gravel should consist of hard, rounded silica stones with an average specific gravity of not less than 2.5. Not more than 1 percent by weight of the material should have a specific gravity of 2.25 or less. Not more than 2 percent by weight of the gravel should consist of thin, flat, or elongated pieces (pieces in which the largest dimension exceeds 5 times the smallest dimension). The gravel should be free from shale, mica, clay, sand, loam, and organic impurities of any kind. The porosity of gravel in any layer should not be less than 35 percent nor more than 45 percent. Gravel should be screened to proper size and uniformly graded within each layer. Not more than 8 percent by weight of any layer should be coarser or finer than the specified limit.

Coarse Garnet

It is recommended that a 3 in. layer of high-density gravel (garnet or ilmenite) be used between the silica gravel bed and the fine media. This coarse, dense layer prevents disruption of the gravel. The specific gravity of the material should be not less than 4.2. The garnet or ilmenite particles in the bottom 1-½ in. layer should be ³/₁₆ in. by 10 mesh, and in the top 1-½ in. layer should be 1 mm in diameter. It is important to note that when the 3 in. coarse garnet layer is not used, another 3 in. layer of ³/₁₆ in. by 10 mesh silica gravel should be added as the top layer in Tables 3-1, 3-2 and 3-3. Otherwise there is apt to be migration of fine media down into the gravel supporting bed.

Gravel Placement

The filter tanks must be thoroughly cleaned before gravel is placed and kept clean throughout the placing operation. Gravel made dirty in any way should be removed and replaced with clean gravel. Place the bottom layer carefully by hand to avoid movement of the underdrain system and to assure free passage of water from the orifices. Complete each gravel layer before the next layer above is started. Workmen should not stand or walk directly on material less than ½ in. in diameter, but rather should place boards to be used as walkways. If different layers of gravel are inadvertently mixed, the mixed gravel must be removed and replaced with new material. The top of each layer should be made perfectly level by matching to a water surface at the proper level in the filter box.

Mixed-Media

For most water purification service, the use of mixed-media (as developed and patented by the Neptune Microfloc Corporation) will provide the optimum filtration efficiency, as discussed previously. A typical performance type specification which will permit use of this media is given in the following paragraph:

The filter beds shall contain a 30 in. total depth of fine media made up of 18 in. of anthracite coal, specific gravity 1.5; 9 in. of silica sand, specific gravity 2.4; and 3 in. of garnet sand, specific gravity 4.2. The relative size of the particles shall be such that hydraulic grading of the material during backwash will result in a filter bed with pore space graded progressively from coarse to fine in the direction of filtration (down). The total surface area of all the grains of fine media in a 30 in. deep bed shall be not less than 2650 square millimeters for each square millimeter of plan area. The number of grains of fine media per square millimeter of plan area shall be not less than 2750. Core samples of the completed backwashed filter bed shall show the number of grains per cubic millimeter given in the following table:

Depth in Bed	Number of Grains per Cubic Millimeter
Top 3 in.	0.59
3–9 in.	0.43
9–15 in.	0.41
15–18 in.	0.40
18–21 in.	0.76
21–24 in.	10.65
24–27 in.	9.85
27–30 in.	16.47

When filtering coagulated and settled wastewater having a turbidity of 10 to 20 JTU (Jackson Turbidity Units) at the rate of 5 gpm/ft², the filter shall produce an effluent having an average turbidity of less than 0.2 JTU and a maximum turbidity of 0.5 JTU even when the filter rate is suddenly changed from 2 to 5 gpm/ft². The initial head loss in a 30 in. deep clean filter shall not exceed 1.6 ft at a filter rate of 5 gpm/ft² and at a water temperature of 20°C. The time required to reach a total head loss of 10 ft (terminal head loss) under the above conditions shall not be less than 12 hr.

There is no one mixed-media design which will be optimum for all water filtration problems. Conley and Hsiung[11] have presented techniques designed to optimize the media selection for any given filtration application. Their work clearly indicates the marked effects that the quantity and quality of floc to be removed can have on media selection (Table 3-4). Pilot tests of various media designs can be more than justified by improved plant performance in most cases.

Certainly, use of three media does not in itself insure superior performance as illustrated by the experiences of Oakley and Cripps.[41] Using the coal, sand, and garnet materials readily available to them resulted in a bed which did not have significant advantages over other filter types. Oakley[42] reports the tri-media bed was made up of 8 in. of 0.7–0.8 mm garnet, 8 in. of 1.2–1.4 mm sand, and 8 in. of 1.4–2.4 mm

**TABLE 3-4   ILLUSTRATIONS OF VARYING MEDIA DESIGN FOR VARIOUS
TYPES OF FLOC REMOVAL[11]**

Type of Application	Garnet size	depth	Silica Sand size	depth	Coal size	depth
Very heavy loading· of fragile floc	−40 + 80	8 in.	−20 + 40	12 in.	−10 + 20	22 in.
Moderate loading of very strong floc	−20 + 40	3 in.	−10 + 20	12 in.	−10 + 16	15 in.
Moderate loading of fragile floc	−40 + 80	3 in.	−20 + 40	9 in.	−10 + 20	8 in.

Size: −40 + 80 = passing No. 40 and retained on No. 80 U.S. sieves.

coal. The authors' experience indicates that this bed was too shallow and too coarse and that a better media selection for the particular application would have been 3 in. of 0.4–0.8 mm garnet, 9 in. of 0.6–0.8 mm sand, and 24 in. of 1–2 mm coal.

One of the key factors in constructing a satisfactory mixed-media bed is the careful control of the size distribution of each component medium. Rarely is the size distribution of commercially available materials adequate for construction of a good mixed-media filter. The common problem is failure to remove excessive amounts of fine materials. These fines can be removed by placing a medium in the filter, backwashing it, draining the filter, and skimming the upper surface. The procedure is repeated until field sieve analyses indicate an adequate particle size distribution has been obtained. A second medium is added and the procedure repeated. The third medium is then added and the entire procedure repeated. Sometimes, 20–30 percent of the materials may have to be skimmed and discarded to achieve the proper particle size distribution.

Dual Media

As compared to mixed-media, the dual media (coal-sand) filter has less resistance to breakthrough because it is made up of coarser particles and has less total surface area of particles. Mixed-media is capable of producing lower finished water turbidities than dual media. These differences are greater and become more pronounced when the difficulty of the filtration application increases. In polishing highly pretreated waters, the differences are not so great, and some designers continue to use coal-sand media. However, it is doubtful that anyone who has observed the two filter types running side by side on the same water would do so.

Typically, coal-sand filters consist of a coarse layer of coal about 18 in. deep above a fine layer of sand about 8 in. thick. Some mixing of coal and sand at their interface is desirable to avoid excessive accumulation of floc which occurs at this point in beds graded to produce well-defined layers of sand and coal. Also, such intermixing reduces the void size in the lower coal layer forcing it to remove floc which otherwise might pass through the coal layer. Typical gradations of sand and coal for use in dual media filters are given in Table 3-5.

TABLE 3-5 TYPICAL COAL AND SAND DISTRIBUTION BY SIEVE SIZE IN DUAL MEDIA BED

Coal Distribution by Sieve Size	
US Sieve No.	*Per Cent Passing Sieve*
4	99–100
6	95–100
14	60–100
16	30–100
18	0–50
20	0–5

Sand Distribution by Sieve Size	
US Sieve No.	*Per Cent Passing Sieve*
20	96–100
30	70–90
40	0–10
50	0–5

"Capping" Sand Filters with Coal

One very easy and inexpensive expedient to improve rapid sand filter performance is to remove about 6 in. of sand from a bed and replace it with 6 in. of anthracite coal. Commonly 0.5 mm sand is capped with 0.9 mm coal. This produces a layered type bed which has only part of the advantages of a dual media bed that has been designed for some intermixing at the interface, but which is superior in performance to a single media (sand or coal) bed. In the October, 1969 *Journal of AWWA*, O. Fred Nelson, manager of the Kenosha, Wisconsin Water Department, describes the results of this modification of sand filters at Kenosha, Sheboygan, and Racine, Wisconsin and at Evanston and Waukegan, Illinois. At Sheboygan, comparisons were made between the coal capped sand filters and sand alone under various raw water conditions, including comparisons made during the algae season. The capped filters were operated at 3 gpm/sf and the sand at 2 gpm/sf. Savings in washwater were reported to be 100 percent. They found that "the more adverse the applied water conditions, relative to algae and floc, the more dramatic are the results obtained with anthrafilt capped filter runs. Good water conditions will give 2 to 1. The worst water conditions may give 10 to 1 improvement in filter runs." "Through the use of capped filters short filter runs can be eliminated." "At Kenosha, residual aluminum tests showed less alum passing through the capped filter than through the sand only filter, except in cases of breakthrough. Bacteriologically, all tests showed safe water at all times." At Racine, it was found that the capped filters removed "more amorphous matter than sand alone." At Evanston, capped filter operation at 4 gpm/sf used less washwater than sand filters.

These benefits of longer filter runs, reduced use of washwater, and better filter effluent quality using capped filters could without question be further extended by replacing all of the fine media with mixed-media. Then, in addition, the mixed-media would eliminate the break-through problem mentioned and would add greatly to the safety and reliability of filter operation. Finished water clarity could be substantially improved through use of a polymer added to the mixed-media bed influent as a filter aid.

Rapid Sand Filters

Virtually all new water purification plants are now being designed to use mixed-media or dual media filters and many existing rapid sand filter plants are being converted to use these materials. However, some single media (sand or coal) filters are being installed in the expansion of existing plants to match existing facilities and to avoid the cost of converting all of the old units.

In the examination of filter sand, the terms "effective size" and "uniformity coefficient" are used. The effective size is the size of the grain in millimeters, such that 10 percent by weight are smaller. The 10 percent by weight corresponds very closely in size to the median size by count, as determined by a size frequency distribution of the total number of particles in a sample. The effective size is a good parameter of the hydraulic characteristics of a sand within certain limits. These limits are usually defined by means of the uniformity coefficient, which is arbitrarily taken as the ratio of the grain size that has 60 percent finer than itself to the size which has 10 percent finer than itself (effective size). This ratio, thus covers the range in size of half the sand.

For practical purposes, the size of sand grains is determined on a weight basis from sieve analysis even though the resulting diameters may be 10–15 percent less than those determined by the count and weigh method, which should be used for more accurate results. The majority of rapid sand filters in use today contain sand with an effective size of 0.35–0.50 mm, although some have sand with an effective size as high as 0.70 mm. The uniformity coefficient is usually not less than 1.3 nor more than 1.7.

Sand passing a 50-mesh (United States series) sieve is generally too fine for use in a rapid sand filter as it stratifies at the surface and shortens filters runs by sealing off the top quite rapidly. Sand retained on a 16-mesh sieve is too coarse to be useful in filtration within the depths normally used in filter plants. Therefore, filter sand usually ranges in size from that passing a 16-mesh to that retained on a 50-mesh sieve. A typical sand specification is given in Table 3-6.

TABLE 3-6 SUGGESTED SIZE SPECIFICATIONS FOR FILTER SAND

| Sieve No. | | | Retained on Sieve percent | |
Series Tyler $\sqrt{2}$	US Series	Opening mm	Minimum	Maximum
65	70	0.208	0	1
48	50	0.295	0	9
35	40	0.417	40	60
28	30	0.589	40	60
20	20	0.833	0	9
14	16	1.168	0	1

Sand filters should have a hydrochloric acid solubility of less than 5 percent when tested in accordance with AWWA Standard B11-53. Sand should have a specific gravity of not less than 2.5, and be clean and well-graded. After placement in the filter, sand should be back-washed 3 times at not less than 30 percent expansion, and then the top ¼ inch of very fine material should be carefully scraped off and discarded.

Crushed anthracite coal may be used in lieu of sand as a fine granular filter medium. It has a specific gravity of 1.5, as compared to 2.65 for silica sand and crushed quartz. Effective sizes of coal up to 0.70 mm are used in filters. Because of the lower specific gravity of coal, only about half the backwashing velocity is needed for equal expansion (not necessarily equal washing) as compared to sand. An anthracite bed of the same effective size as a sand bed has a greater bed porosity. Anthracite coal filter media should be clean and free from long, thin, or scaly pieces, with a hardness of 2.0 to 3.55 on the MOH scale and a specific gravity not less than 1.5.

Fine Media Placement

Media should be transported and placed carefully to prevent contamination of any sort; and media made dirty before or after placement should be replaced with clean media.

The fine media for mixed media are placed in the bed in the order of their respective specific gravities, the highest specific gravity material (garnet) being placed first and leveled. Then the sand is placed and leveled. The bed is then backwashed 3 times, and ½ in. of the surface fines removed by scraping. Finally the coal is to be placed and finished off smooth to the proper elevation. The bed should again be backwashed 3 times and ½ in. of the surface fines removed by scraping.

Filter Backwashing

During the service cycle of filter operation particulate matter removed from the applied water accumulates on the surface of the grains of fine media and in the pore spaces between grains. With continued operation of a filter the materials removed from the water and stored within the bed reduce the porosity of the bed. This has two effects on filter operation; it increases the head loss through the filter and increases the shearing stresses on the accumulated floc. Eventually the total hydraulic head loss may approach or equal the head necessary to provide the desired flow rate through the filter, or there may be a leakage or breakthrough of floc particles into the filter effluent. Just prior to either of these potential occurrences, the filter should be removed from service for cleaning. As mentioned previously, in the old slow sand filters the arrangement of sand particles is fine-to-coarse in the direction of filtration (down), and most of the impurities removed from the water collect on the top surface of the bed and the bed can be cleaned by mechanically scraping the surface and removing about one half inch of sand and floc. In rapid sand filters, there is somewhat deeper penetration of particulates into the bed because of the coarser media used and the higher flow rates employed. However most of the materials are stored in the top 8 in. of a rapid sand filter bed. In dual media and mixed-media beds, flow is stored throughout the bed depth to within a few inches of the bottom of the fine media. Rapid sand, dual media, and mixed-media filters are cleaned by hydraulic backwashing (upflow) with potable water. Thorough cleaning of the bed makes it advisable in the case of single media filters and mandatory in the case of dual or mixed media filters to use auxiliary scour or so-called "surface wash" devices before or during the backwash cycle. Backwash flow rates of 15 to 20 gpm/sf should be provided. A 20 to 50 percent expansion of the filter bed is usually adequate to suspend the bottom grains. The optimum rate of wash water application is a direct function of water temperature, as expansion of the bed varies inversely with viscosity of the wash water. The time required for complete washing varies from 3–15 min. Following the washing process, water should be filtered to waste until the turbidity drops to an acceptable value. Filter-to-waste outlets should be through an air gap-to-waste drain which may require from 2–20 minutes, depending on pretreatment and type of filter. This practice was discontinued for many years, but modern recording turbidimeters have shown that this operation is valuable in the production of a high-quality water. Operating the washed filter at a slow rate at the start of a filter run may accomplish the same purpose. A recording turbidimeter for continuous monitoring of the effluent from each individual filter unit is of great value in controlling this operation at the start of a run as well as in predicting or detecting filter break-through at the end of a run.

Filters can be seriously damaged by slugs of air introduced during filter backwashing. The supporting gravel can be overturned and mixed with the fine media, which requires removal and replacement of all media for proper repair. Air can be unintentionally introduced to the bottom of the filter in a number of ways. If a vertical pump is used for the backwash supply, air may collect in the vertical pump column between backwashings. The air can be eliminated without harm by starting the pump against a closed discharge valve and bleeding the air out from behind the valve through a pressure air release valve. The pressure air release valve must have sufficient capacity to discharge the accumulated air in a few seconds.

Also, air or dissolved oxygen, released from the water on standing and warming in the wash water supply piping, may accumulate at high points in the piping and be swept into the filter underdrains by the inrushing wash water. This can be avoided by placing a pressure air release valve at the high point in the line, and providing a ½ in. pressure water connection to the wash water supply header to keep the line full of water and to expel the air.

The entry for wash water into the filter bottom must be designed to dissipate the velocity head of the wash water in such a manner that uniform distribution of wash water is obtained. Lack of attention to this important design factor has often led to difficult and expensive alterations and repairs to filters for correction.

Horizontal centrifugal pumps may be used to supply water from the treated water storage reservoir for filter backwashing provided they are located where positive suction head is available, or are provided with an adequate priming system. Otherwise, a vertical pump unit may be suspended in the clear-well. The pumps should be sized to supply at least 15 gpm/ft² of filter area to be washed, and should be installed with an adequate pressure air release valve, a nonslam check valve, and a throttling valve in the discharge. Single wash water supply pumps are often installed, but consideration should be given to provisions for standby service in areas where power may be interrupted for several hours at a time, or where pump or motor repairs are not quickly available.

Wash water may also be supplied by gravity flow from a storage tank located above the top of the filter boxes. Wash water supply tanks usually have a minimum capacity equal to a 7 min wash for 1 filter unit, but may be larger. The bottom of the tank must be high enough above the filter wash troughs to supply water at the rate required for backwashing as determined by a hydraulic analysis of the wash water system. This distance is usually at least 10 ft, but more often is 25 ft or greater. Wash water tanks should be equipped with an overflow line, and a vent for release and admission of air above the high water level. Wash water tanks are often filled by means of a small electric driven pump equipped with a high-water level cutoff. They may also be filled from the high-service pump discharge line through an altitude or float valve, but this is usually wasteful of power. Wash water tanks may be constructed of steel or concrete. Provisions should be made for shutdown of the tanks for maintenance and repair.

The wash water supply line must be equipped with provisions for accurately (within ± 5 percent) measuring and regulating the rate of wash water flow. The rate of flow indicator should be visible to the operator at the point at which the rate of flow is regulated.

The use of a high pressure (above 15 psi) source of filter backwash water through a pressure-reducing valve is not advised. Numerous failures of systems using pressure reducing valves have so thoroughly upset and mixed the supporting gravel and fine media that these materials had to be completely removed from the filter and replaced with new media.

Wash Water Troughs

To equalize the head on the underdrainage system during backwashing of the filter, and thus to aid in uniform distribution of the wash water, a system of weirs and troughs is ordinarily used at the top of the filter to collect the backwash water after it emerges from the sand and to conduct it to the wash water gullet or drain. The bottom of the trough should be above the top of the expanded sand to prevent possible loss of sand during backwashing. The clear horizontal distance between troughs is usually 5–7 ft. The cross section of the trough depends somewhat on the material of construction, but the semicircle bottom is a good design. Troughs may be made of concrete, fiberglass reinforced plastic, or other structurally adequate and corrosion-resistant materials. The dimensions of a filter trough may be determined by use of the following equation:

$$Q = 2.49 \ bh^{3/2}$$

in which, Q is rate of discharge in ft³/sec, b is width of trough in ft, and h is maximum water depth in trough in ft. Some freeboard should be allowed to prevent flooding of wash water troughs and uneven distribution of wash water.

The most popular material for construction of wash water troughs is reinforced concrete. Ordinarily concrete troughs are precast on the job in an inverted position, and then placed in position after forms have been removed. Wash water troughs may also be factory fabricated from steel, fiberglass, or transite. Fiberglass reinforced plastic troughs may be purchased from F. B. Leopold Co., Inc., Zelienople, Pa., Fischer & Porter, Warminster, Pa., or other manufacturers. Transite wash troughs are available from Filtration Equipment Corp., Rochester, N. Y. and other suppliers.

If the filter troughs are to properly serve their purpose, after placement their weir edges must be honed to an absolutely smooth and perfectly level edge as determined by matching the finished edges of all troughs in a single filter to a still water surface at the desired overflow elevation.

Filter Agitators

Present practice in the U. S. leans heavily toward installation of the essential (but misnamed) surface wash on all new filters. Auxiliary scour or filter agitation better describe the function of this device, as it aids in cleaning much more than the filter surface. Rotary surface washers are the most common, but fixed jets are also used successfully. In the past, air scour and mechanical scour have not been used extensively in the United States, but recent developments in this area may change this in the future.

Adequate surface wash improves filter cleaning and prevents mudball formation and filter cracking. Rotary surface wash equipment consists of 2 arms on a fixed swivel supported from the wash water troughs about 2 in. above the surface of the unexpanded filter media. The arms are fitted with a series of nozzles, and revolve because of the water jet reaction. Water pressures of 40–100 psi are required for the operation of the rotary surface wash depending upon the diameter of the arms. The volume required is about 0.5 gpm/ft². Surface washers are usually started about 1 min in advance of the normal backwash flow, and turned off a minute or so before the end of the backwash period. Rotary surface wash equipment can be purchased from Leopold. They can supply either the Palmer or Stuart rotary agitators. It is recommended

(a)

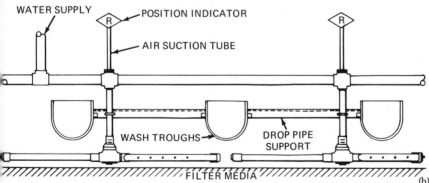

(b)

Fig. 3-6   *The Roberts XL-500 rotary Agitator.*

   (a)   Here the sweep arms are held stationary to illustrate the violent action of the sprays using an air/water mixture. At this point the control ports in the rotary joint are still closed as the end nozzle passes the wall of the filter.

   (b)   Elevation Drawing of a typical installation showing relationship of the stator supports to wash troughs.

that the nozzles be equipped with Stuart Flex-I-Jet rubber caps which act to prevent entry of fine filter media and plugging of the nozzles.

A new concept in rotary agitator design has been developed by Roberts Filter Mfg. Co., Darby, Pa. (see Figure 3-6). This agitator utilizes the aspirator effect of the agitator water supply to introduce a controlled amount of air into the agitator arms. The air-water jet spray provides greatly improved scour and separation of accumulated floc. The agitators also include a rotary valve which cuts off the water supply to the end sprays during the instant that the arms are closest to the filter walls. This reduces erosion of the concrete at this point which is sometimes a problem with conventional rotary surface washers.

Baylis has designed a fixed-jet surface wash system consisting of a grid of distributing pipes extending to within a couple of inches of the top surface of the bed. Nozzles with five ¼ in. holes are spaced at about 24–30 in. centers each way. The required flow is about 2 gpm/ft² at a head of 20–60 ft.

Surface wash piping is a direct cross connection between filtered and unfiltered water, so that the normal practice is to bring the surface wash header into the filter over the top of the filter box and to fit it with a vacuum breaker and a check valve at the high point to prevent back-syphonage. A single vacuum breaker on the surface wash header will suffice. An alternate is to use settled water through a separate surface wash pump. Valve positions for various cycles of filter operation are shown in Table 3-7.

TABLE 3-7   VALVE POSITIONS DURING VARIOUS TREATMENT OPERATIONS

| | Valve Position | | |
Valve	Filtering	Backwashing	Filtering to Waste
Influent	Open	Closed	Open
Effluent	Open	Closed	Closed
Filter-to-waste	Closed	Closed	Open
Wash water supply	Closed	Open	Closed
Wash water drain	Closed	Open	Closed
Surface wash	Closed	Open	Closed

Filter Rate-of-Flow Controllers

Ordinarily filters are equipped with a means of controlling the rate of flow through each bed. By controlling either the influent or the effluent flow it is possible to divide the flow among the filters, to limit the maximum flow through any unit, and to prevent sudden flow surges. Rapid changes in rate of flow through a filter are undesirable because they literally shake the floc particles down through the bed and into the effluent.

Filter rate of flow controllers consist of (1) a flow measuring device such as a Venturi tube or a Dall tube, (2) a throttling valve such as a rubber seated butterfly valve with hydraulic or pneumatic actuator, and (3) a valve stop or positioning device. Auxiliary or overriding control of the throttling valve may be provided to maintain a predetermined minimum water level above the filter media and a maximum water level in the receiving clearwell. The flow measuring device must be provided with the necessary straight pipe runs on either side if it is to provide the necessary accuracy. Controllers can be equipped with gauges or charts to indicate or to record rate of flow through each filter, and to summarize total flow.

Rate controllers may be set to operate at a constant fixed rate, which has several advantages from the standpoint of water quality. This means that flow through the plant must be nearly constant, or that filter units must be thrown in or out of service to meet fluctuating demands. Another method of operation is to use a variable-rate controller with a fixed-maximum setting. This permits operation of all filters at the minimum possible rate all of the time, which also has advantages.

Loss of Head

The loss of head through a filter provides valuable information about the condition of the bed and its proper operation. An increase in the initial loss of head for successive runs over a period of time may indicate clogging of the underdrains or gravel, the need for auxiliary scour, or insufficient washing of the beds. The rate of headloss increase during a run yields considerable information concerning the efficiency both of pretreatment and filtration.

The determination of headloss through a filter is a very simple matter, since it involves only the measurement of the relative water levels on either side of the filter. The simplest form of headloss device for gravity filters is one made up of 2 transparent tubes installed side by side in the pipe gallery with a gauge board between, graduated in feet. One tube is connected to the filter or filter influent line, the other to the effluent line. The headloss is the observed difference in water levels. More sophisticated methods for measuring the difference in water levels include float-and differential-pressure-cell-actuated indicating and recording devices. The headloss may be indicated and recorded at some remote point if desired. Headloss equipment may be used in connection with control systems for automatic backwashing of filters as one means of initiating the backwash cycle when the headloss reaches some preset maximum value.

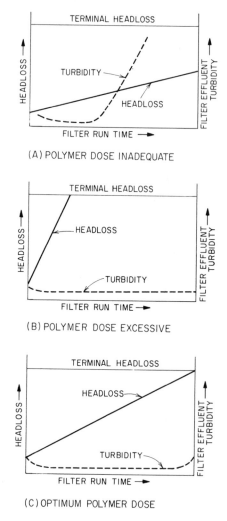

Fig 3-7 Effects of polymers as filtration aids.

Use of Polymers as Filtration Aids

Polymers are high molecular weight, water soluble compounds which can be used as primary coagulants, settling aids, or filtration aids. They may be cationic, anionic, or nonionic in charge. Generally, the doses required as a filtration aid are less than 0.1 mg/l. Used as a filtration aid, polymer is added to increase the strength of the chemical floc and to control the depth of penetration of floc into the filter. For maximum effectiveness as a filtration aid, the polymer should be added directly to the filter influent and not in an upstream settling

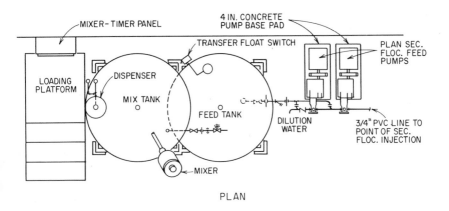

PLAN

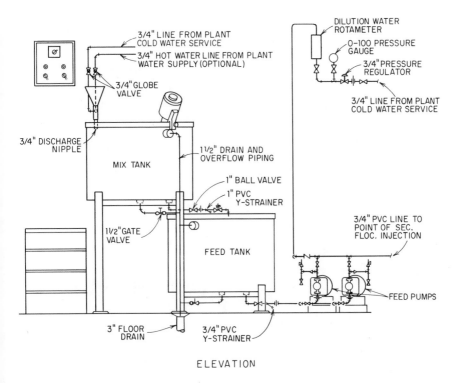

ELEVATION

Fig. 3-8 Polymer mixing and feed equipment. *(Courtesy Neptune Microfloc, Inc.)*

basin or flocculator. However, if polymers are used upstream as settling aids, it may not be necessary to add any additional polymer as filtration aid.

Figure 3-7 illustrates the effects of polymers as filter aids. The conditions represented by Figure 3-7 illustrate the results of a fragile floc shearing and then penetrating the filter causing a premature termination of its run due to breakthrough of excessively high effluent turbidity. If the polymer dose is too high (Figure 3-7B) the floc is too strong to permit penetration into the filter causing a rapid buildup of head loss in the upper portion of the filter and a premature termination due to excessive headloss. The optimum polymer dose will permit the terminal headloss to be reached simultaneously with the first sign of increasing filter effluent turbidity (Figure 3-7C).

Many polymers are delivered in a dry form. They are not easily dissolved and special polymer mixing and feeding equipment is required. Typical equipment is shown in Figure 3-8 and 3-9. Many polymers are biodegradable and cannot be stored in dilute solution for more than a few days without suffering significant degradation and loss of strength. Further data on the control of polymer dosage to aid filtration are given in chapters which follow.

Fig. 3-9   Typical polymer feed station. *(Courtesy Neptune Microfloc, Inc.)*

Monitoring Filter Effluent Turbidity

Inexpensive, reliable turbidimeters are available to continuously monitor the quality of filter effluent. They are described in detail in a later chapter. Should the turbidity exceed the desired level, a signal from the turbidimeter can be used to sound an alarm or to initiate the backwash program. Recording of the turbidimeter output provides a continuous record of filter performance.

Pilot Filters

A small filter receiving raw water dosed with the treatment chemicals is often operated in parallel with the plant filters. The pilot filter directly measures the filtrability of the water under actual plant operating conditions. By fitting the pilot filter with a recording turbidimeter, chemical dosages can be adjusted to obtain the desired quality of filter effluent before water in the full scale plant reaches the filters. Detailed descriptions of pilot filters and their use are contained in a later chapter.

Control of Filter Operation

Until about 1960, most filter plants depended primarily upon pretreatment facilities to produce filter influent of low turbidity and low filter rates (3 gpm/sf or less) to produce a satisfactory filter effluent. Composite or grab samples of filtered water were examined for turbidity or bacterial content. The results of these tests were used to adjust pretreatment and for record purposes, rather than for direct control of filter operation. Cotton plug filters were introduced by Baylis and used by others to evaluate filter performance based on long-term operation. Since 1960, three major developments have greatly increased the degree of control over the efficiency and reliability of filter operation: (1) the use of 0.01 to 0.05 mg/l of a polymer to control floc strength, depth of floc penetration into the bed, and effluent turbidity, (2) the use of continuous recording turbidimeters having a sensitivity of 0.01 JTU, and (3) the use of a control bed or pilot filter to anticipate pretreatment requirements for optimum filter performance. These three major advances allow a degree of control and direct adjustment of filter operation not previously attainable. It should be pointed out that the use of polymer as a filter aid is not suited to single media, fine-to-coarse filters as they are sealed off quite rapidly at the surface by such application to the filter influent.

Filter runs are usually terminated when either the head loss reaches a predetermined value, the filter effluent turbidity exceeds the desired maximum, or a certain amount of time passes. Each of these events is adaptable to instrumentation which can be used to signal the need for backwashing or to trigger a fully automated backwash system. Automatic control of filter backwashing may be provided by an automatic sequencing circuit (step switch) which is interlocked so that the necessary prerequisites for each step are completed prior to proceeding to the next step. At the receipt of a backwash start signal, the following events will occur in the sequence listed in this illustrative program. Filter influent and effluent valves close. Any chemical feed stops to the filter being backwashed. Plant chemical feeds adjust to the new plant flow rate to maintain proper chemical feed to the filters still in service. The waste valve starts to open. When the waste valve reaches the fully open position and actuates a limit switch, the surface wash pump starts and the surface wash valves open. Surface wash flow to waste continues for a period of time adjustable up to 10 min.

At the end of the initial surface wash period, usually 1 to 2 min, the main backwash valve opens. Backwash and surface wash both continue for a period of time, usually 6 to 7 min, adjustable up to 30 min. Backwash flow rate is indicated on a controller and is controlled automatically to a manual set point. At the end of the combined wash periods, the surface wash valves close and surface wash pump stops. Backwash continues without surface wash for a period, usually 1 to 2 min, adjustable up to 30 min. At the completion of the backwash period, the backwash valve has closed, the waste valve starts to close. When the latter valve has closed, influent and effluent waste valves open and the bed filters to waste for a period of time, usually 3 to 7 min, adjustable up to 30 min. The backwash delay timer resets and begins a new timing cycle adjustable up to 12 hr. The bed selector switch steps to the next filter. Chemical feed to the clean filter is reestablished. At the end of the filter-to-waste period, the effluent waste valve closes and the effluent valve opens to restore the cleaned bed to normal filter service. Provision should be made for optional manual operation of all automatic features.

It may be desirable to alarm certain functions which affect filter operation on a conveniently located annunciator panel. These alarm functions include high turbidity, high head loss, low plant flow, low backwash flow rate, and excessive length of backwash.

Filter Piping, Valves, and Conduits

Cast-iron pipe and fittings and coal-tar enamel-lined welded steel pipe and fittings are the most widely used materials for filter piping. The layout of filter piping must include consideration of ease of valve removal for repair and easy access for maintenance. Flexible pipe joints should be provided at all structure walls to prevent pipeline breaks due to differential settlement. The use of steel pipe can reduce flexible joint requirements. Color coding of the filter piping is a valuable operating aid. The filter piping is usually designed for the flows and velocities shown in Table 3-8.

TABLE 3-8 FILTER PIPING DESIGN FLOWS AND VELOCITIES

Description	Velocity (fps)	Maximum flow, gpm/sf of filter area
Influent	1–4	8–12
Effluent	3–6	8–12
Wash water supply	5–10	15–25
Backwash waste	3–8	15–25
Filter to waste	6–12	4–8

The rubber-seated, pneumatically actuated and operated butterfly valve has almost entirely replaced the hydraulically actuated and operated gate valves that were formerly used extensively as filter valves. The butterfly valve is smaller, lighter, easier to install, better for throttling services, and can be installed and operated in any position. The valves should be factory equipped with the desired valve stops, limit switches, and position indicators because field mounting of these devices is often unsatisfactory.

Each filter unit, except split beds, should have 6 valves for its proper operation: influent, effluent, wash water supply, wash water drain, surface wash, and filter to waste. The positions of these valves during the three cycles of filter operation were given in Table 3-7. The filter influent should enter the filter box so that the velocity of the incoming water does not disrupt the surface of the fine media. This is often done by directing the influent stream against the gullet wall, thus dissipating the velocity head within the gullet. It can also be done by locating the influent pipe below the top of the filter troughs so that the water enters the filter through the troughs. A further precaution is to install an influent valve with throttling control for use in refilling the beds slowly. The filter-to-waste connection to the filter should have positive air gap protection against back siphonage from the drain to the filter bottom. The filter-to-waste, effluent, and wash water supply lines usually are manifolded for common connection to the filter underdrain system.

In the design of pipe galleries, reinforced concrete flumes and box conduits and concrete encased concrete pipe may be used for wash water drains or other service when located adjacent to the pipe gallery floor, but should not be installed overhead because of difficulties with cracks and leaks. Invariably, pipe galleries with overhead concrete conduits are drippy, damp, unsightly places with a humid atmosphere which discourages good housekeeping by making it difficult to maintain. Rather, pipe galleries should be provided with positive drainage, good ventilation, plenty of light, and dehumidification equipment if required by the prevailing climate. Filter influent and effluent lines

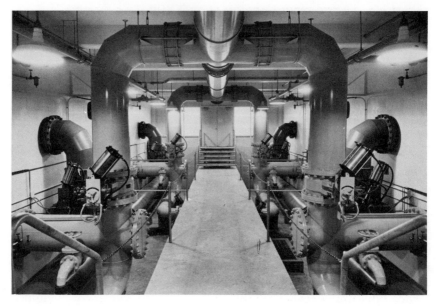

Fig. 3-10   Filter pipe gallery at the Richland, Washington water treatment plant.

should be provided with sample taps. Figure 3-10 is a photograph of the filter pipe gallery at the Richland, Washington Water Treatment Plant.

Filter Problems and Their Solutions

Problems in filter operation and performance can arise either from poor design or poor operation. However, during the past 10 years the tremendous advances in engineering design of filters and filter controls and appurtenances have made water filtration an inherently stable, extremely efficient, and highly reliable unit treatment process. With proper design and good operation, all the problems of the past are easily solved.

Some potential filter problems are listed below:

1. Surface clogging and cracking.
2. Short runs due to rapid increases in headloss.
3. Short runs due to floc breakthrough and high effluent turbidity.
4. Variations in effluent quality with changes in applied water flow rate or quality.
5. Gravel displacement or mounding.
6. Mud balls.

7. Growth of filter grains, bed shrinkage, and media pulling away from sidewalls.
8. Sand leakage.
9. Loss of media.
10. Negative head and air binding.
Some solutions to these problems are suggested below:

1. Surface clogging and cracking are usually caused by rapid accumulations of solids on the top surface of the fine media. This is not a problem in dual or mixed-media filters because of the greater porosity of their top surface as compared to sand. Also, when a filter aid is used with dual or mixed-media, the dosage can be reduced as necessary (or eliminated) which allows particulates to penetrate deeper into the bed. In other words, regulation of the polymer dosage to the filter influent gives some control over the effective porosity of the filter so as to accommodate changes in incoming floc characteristics.

2. Short runs due to rapid increases in headloss are related to the problem just discussed in that dual and mixed-media beds collect particulates throughout the depth of the bed rather than mostly at the surface of the bed, as with a sand or other surface type filter, and are much less susceptible to this problem. Also the flexibility provided by use of a polymer as a filter aid allows control of the rate of headloss buildup by dosage changes.

3. Short filter runs due to floc breakthrough can be avoided by using mixed-media. As mentioned earlier, this is one very important point of superiority of mixed-media over dual media and sand filters. It arises because of the much greater surface area of the grains in a mixed-media filter as compared to sand or dual media. The finest media in a mixed-media bed is 40 to 80 mesh (10 percent of total bed), 40 to 50 mesh (9 percent of total) in a sand bed, and 40 to 50 mesh (5 percent of total) in a dual media filter. The finest media (garnet) in a mixed-media bed has the advantage not only of being finer but also in being located at the very bottom of the filter where the applied load is lightest and where it can serve its intended purpose as a polishing agent. Floc storage depths above the finest media in various typical beds are as follows: sand, 0 in.; coal-sand, 18 in.; and mixed-media, 27 in. A further advantage of mixed media in this regard lies in the greater total number of media particles contained in an equal volume of bed. This tremendously increases the number of opportunities for contact between media and colloids in the water which greatly enhances removal of these colloids. It is these superior properties of mixed-media in resistance to leakage of particulates and much greater removal of colloids that make it questionable to use single or dual media beds under any circumstances.

4. Variation in effluent quality with changes in applied water are

much less in mixed-media beds which are much less affected either by changes in flow or influent quality than either sand or dual media. Again this relates to the greater total surface area of the fine media, the greater total number of fine media particles, and the smaller size of pore openings at the bottom of the filter bed.

5. Gravel displacement or mounding can be eliminated by use of the 3 in. layer of coarse garnet or ilmenite between fine media and gravel supporting bed as previously recommended, and by limiting the total flow and head of water available for backwash (that is, not drawing wash water from a high pressure source through a pressure reducing valve which could fail).

6. Mud ball formation can be eliminated by providing an adequate backwash flow rate (up to 20 gpm/sf), and a properly designed system for auxiliary scour (surface wash). The successful use of beds employing filter aids and in-depth filtration is dependent on provision and operation of a good system of auxiliary scour.

7. Growth of filter grains, bed shrinkage, and media pulling away from filter sidewalls are related problems. Again, the provision and use of adequate backwash facilities including surface wash are the keys. It is the compressibility of filter grains which are heavily coated with materials filtered, deposited, or absorbed from the water which is the root of these difficulties. With one exception to be discussed later, these problems can be avoided by proper backwashing. It should be mentioned that this growth of particles refers to a macroscopic increase in size and not to the development of a microscopic film of polymer and other chemicals which result from proper use of a filter aid and which is beneficial in adsorption and retention of particulates for the period of a single operational cycle in the use of dual and mixed-media beds and which is not nearly thick enough to be a problem by increasing the compressibility of the bed. Actually in the one exception alluded to earlier, which is calcium carbonate deposition on filter grains, an alum or polymer film on filter grains may actually be an aid in reducing the adherence of calcium carbonate and facilitating its removal during backwashing. However, in any case, in filtering lime softened waters it is important to adjust the pH of filter influent by addition of carbon dioxide or acid to a level at which calcium carbonate deposition does not occur.

8. Sand leakage can be prevented by using the coarse garnet layer between the fine media and gravel supporting bed as recommended earlier, as this prevents the downward migration and escape of fines.

9. Loss of media, particularly of coal during backwashing, is one problem for which there is no complete solution. Losses can be reduced by increasing the distance between the top of the expanded bed during maximum backwash flows and the wash water troughs. It also can be

helped by cutting off air wash or auxiliary scour 1 or 2 min before the end of the main backwash.

10. Negative head and air binding can be avoided in most cases, but there may be a few extreme situations, usually of short duration, where they cannot be entirely eliminated. In any case it is a good idea to provide a water depth of at least 5 ft above the surface of the unexpanded filter bed. The more depth the better, so far as negative head and air binding are concerned.

When filter influent water contains dissolved oxygen at or near saturation levels, and when pressure is reduced to less than atmospheric below the surface of the fine media by siphon action, the oxygen comes out of solution and gas bubbles are released. They may accumulate within the bed and tremendously increase the resistance to flow or head loss. When flow through a filter is stopped and the water level is lowered in preparation for backwashing, bed pressures are reduced and more oxygen is released. Even further release of bubbles occurs during backwash which may lead to loss of media in the waste backwash water when bubbles adhere to coal or sand and carry them into the wash water troughs. More frequent filter backwashing may alleviate the problem to some extent as there is then less time for bubbles to accumulate, but when the problem is acute, as it may be in the spring when surface water is warming (and oxygen solubility is decreasing at the higher temperatures) it is a problem which under the very worst conditions may only be endured and not solved. Maintaining maximum water depths above the beds and frequent backwashing may help but may not completely eliminate the difficulties.

Filter Loading

The fact that coarse-to-fine, in-depth filters as exemplified by mixed-media beds can store such large quantities of particulates which have been removed from the water without excessive head loss or floc breakthrough opens up possibilities of several plant flow sheets other than the standard approach of full rapid mixing, flocculation, and settling as treatment prior to filtration. One of these is direct filtration of low turbidity coagulated waters without the need for sedimentation. This has been done in several applications. In general, direct filtration can be applied when the average raw water turbidity is in the range of 25 to 50 JTU. Occasional turbidity peaks of 100 JTU can also be handled without difficulty. Another flow sheet for low turbidity waters provides a contact basin with 30 to 60 min holding time but without installation of equipment for the continuous collection and removal of sludge. The addition of the contact basin increases the potential range of turbidities which can be processed to about 20 to 100 JTU on the average with tolerable peaks as high as 200 to 1000 JTU. The presentation of design and operating data from a few plants which have been in service several years using such modified approaches follow in a later chapter.

Existing Plants-Expansion and Conversion

Because mixed-media filters operate more efficiently, safely, and reliably at 5 gpm/sf than do conventional rapid sand filters at only 2 gpm/sf, there obviously is great potential for expanding the capacity of existing plants at least up to double with only the nominal expense of replacing sand with mixed-media. This, of course, has been done in a great many instances. Because of the ability of mixed-media filters to remove and store solids from high turbidity waters, often it is not necessary to add settling basin capacity in plant expansion. In other cases this must be done, or, as an alternative, settling tubes may be installed in existing basins, since this change will allow increasing basin throughput without loss of settling efficiency.

In an article in the 1968 AWWA Journal, page 699, titled, "Necessary Modifications for High Rate Filtration," Mark C. Culbreath outlines rather completely the items to be considered in expanding existing plants by conversion to mixed-media filtration. Most of these necessary modifications have to do with low service pump capacity, the hydraulic capacity of flumes, pipelines, and conduits, the possible need for a new intermediate pumping lift, increased chemical feed capacity, and additional high service pumps.

To date (1973) more than 150 existing rapid sand filter plants have been converted to mixed-media in order to increase plant capacity. They have also gained improvement in water quality and greater control over plant operations. An example of a converted plant is the one at Winnetka, Illinois. This 6 mgd plant is owned and operated by the village of Winnetka. It was of traditional design, utilizing flocculation, settling, and 8 rapid sand filters to provide clarification of Lake Michigan water for municipal use. Early in 1967, 4 of these filters were converted to mixed-media to increase the capacity of the plant to 11 mgd. Larger filter rate-of-flow controllers were also installed. No other changes in plant equipment were required. The capacity of the converted filters was increased from 2 to 5 gpm/sf by installation of the mixed-media in place of the sand and the larger rate controllers. A coagulant control center was also installed. It is used to determine alum feed needs and to continuously monitor the turbidity of the filtered water entering the clearwell. The Winnetka plant reports a 50 percent savings in backwash water with the use of the mixed-media filters. The expansion of this plant from 6 to 11 mgd was accomplished at a cost of less than $90,000 as compared to an estimated $650,000 for installation of a conventional treatment system for the same additional capacity (see Chapter 10).

Application of Mixed-Media Technology to New Plants

During the past 10 years, tremendous strides have been made in production of potable water by application of mixed-media technology which includes: (1) the addition of a filter aid such as alum, activated silica, or a polymer to filter influent, (2) use of pilot control filters, (3) continuous monitoring of individual filter effluent turbidities, and (4) use of mixed-media supported on a coarse bed of garnet. The original goal of Pitman and Conley, which was to develop methods for consistently producing water of extremely low turbidity, has been realized. While they sought to do this in order to improve the quality of water for processing radioactive materials at the Hanford Works, this improvement in quality has had an even greater significance in the contribution it has made to the protection of public health. Hudson has demonstrated rather conclusively that the ability to destroy virus in water is directly related to reducing the turbidity of the water to be disinfected to 0.2 JTU or less prior to final chlorination. The reason for this greatly improved viricidal action of chlorine at low turbidities is believed to be that there is no opportunity for encapsulation of virus in particulate matter, thus all virus can be contacted by the chlorine and destroyed. Before mixed-media filtration, only a very few of the best operated plants in the country could consistently maintain the low turbidities required to assure complete virus elimination. Now it has become a simple routine operational task accomplished by a great many filter plants throughout the country.

The new technology provides for the first time direct, firm, and almost foolproof control over the coagulation-filtration process. It has progressed from the point of being viewed with considerable skepticism as a radical innovation in the early 1960s to the rather generally new standard of the water works industry in 1973. As shown by Figure 3-11, there are now more than 250 mixed-media filter plants producing more than 1 billion gallons per day of water for municipalities and industries.

In addition to greatly simplifying the process of obtaining high quality water, the new technology has other advantages. It substantially reduces space requirements for plant facilities. It cuts the cost of plant expansion or new plant construction by 30 percent or more, and also reduces plant operating costs through the use of less chemicals and wash water. It greatly improves the flexibility and reliability of the coagulation-filtration process.

The facts that the locations of these improvements are so widely

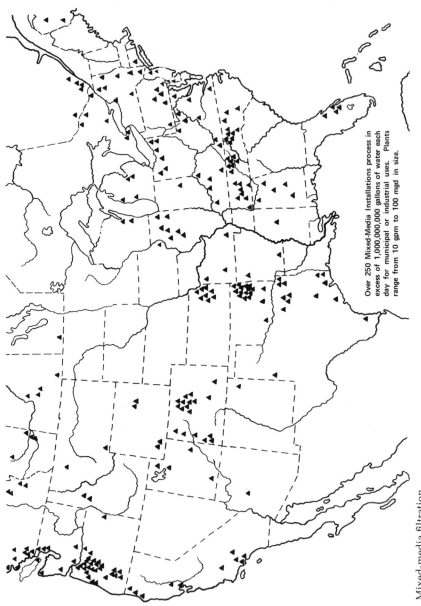

Over 250 Mixed-Media Installations process in excess of 1,000,000,000 gallons of water each day for municipal or industrial uses. Plants range from 10 gpm to 100 mgd in size.

Fig. 3-11 Mixed-media filtration.

scattered and that the changes have taken place over a period of 10 years make it difficult to realize that such far-reaching advances have been made, but indeed they have. Chapter 10 presents specific examples of applications to new plants.

OTHER TYPES OF FILTERS

Diatomite Filters

Diatomite filters consist of a layer of diatomaceous earth about ⅛ in. thick supported on a septum or filter element. The thin precoat layer of diatomaceous earth is subject to cracking and must be supplemented by a continuous body feed of diatomite. The problems inherent in maintaining a perfect film of diatomaceous earth between filtered and unfiltered water have restricted the use of diatomite filters for municipal purposes, except under certain favorable conditions.

The present status of the diatomite filter and its potential uses are well-summarized in the Third Edition (1971) of *Water Quality and Treatment* as published by the AWWA, as quoted in the paragraphs which follow:

> The diatomite filter is an acceptable tool in the waterworks industry, but is only a step in the water-treatment process, the details of which must be predicated on the characteristics of the raw water and what must be done to condition the water prior to filtration. Present municipal experience relates principally to turbidity removal from surface waters, but experience is growing with installations designed for iron and manganese removal from groundwaters and with filtration of lime-soda ash-softened water and coagulated surface waters. Municipal experience with turbidity removal has been principally where the actual turbidity is relatively low and the bacteriological quality is good, thus little pretreatment is provided. There is no common agreement on the upper limits of turbidity that can be handled without pretreatment. There has been some extension of the use of diatomite filters in some of the areas (i.e., marginally, moderately, and grossly polluted surface waters) which have been of particular concern to regulatory public health personnel. No one should recommend diatomite filtration in the near future for filtration of a grossly or even moderately polluted supply. However, appropriate prior conditioning of such waters by one of several available means, including coagulation and settling or chemical treatment of the filter medium or water to improve water filtrability, will undoubtedly lead to wider application of diatomite filters. Currently, effective water conditioning coupled with filtration through carefully engineered filters, an appropriate filter aid body feed system, and with adequate disinfection should provide a substantial margin of safety for marginally polluted waters and even sometimes for moderately polluted waters.

An advantage of a diatomite filtration plant for potable water is the lower first cost of such a plant. On waters containing low suspended solids, the diatomite filter installation cost should be somewhat lower than the cost of a conventional rapid-sand filtration plant. Diatomite filters will thus find application in potable water treatment under the following conditions:

1. In cases where the diatomite plant will be found to produce water at a lower total cost than any practical alternative.

2. In cases where financial capacity is tightly circumscribed, the lower first cost of a diatomite filter installation may be the major factor in the final choice of plant.

3. For emergency or standby service at locations experiencing large seasonal variations in water demand, the lower first cost of the diatomite filter may prove to be more economical.

Microstraining

The process of microstraining incorporates a specially woven stainless steel wire cloth mounted on the periphery of a revolving drum fitted with continuous backwashing arrangements. In operation, the drum is submerged in the flowing water to approximately two thirds of its depth. Raw water enters through the upstream open end of the drum and flows radially outwards through the microfabric, leaving behind its suspended solids content. These intercepted solids are carried upwards on the inside of the fabric under a row of washwater jets, from where they are flushed into a receiving hopper mounted on the hollow axle of the drum.

The wire cloths used in microstraining have apertures of 60, 35, and 23 μ with 60,000, 80,000 and 165,000 openings/in.[2] The fabrics may be woven so that the apertures are "three-dimensional" to provide high porosity, with durability and ease of backwashing. The flow capacity of a microstrainer depends on the rate of clogging of the fabric, drum speed, area of submergence, and head loss. These factors are combined in an exponential equation to obtain required plant size.

A portable testing kit has been developed to determine the filtration rate of any water source. Water under test is held in a reservoir and the time of discharge through a ½ in. diameter disc of micro fabric determines the "Filterability Index." Using this set, it is possible to specify the size of installation and operating efficiency for any project.

Microstraining has been used for removal of troublesome algae from lake waters upstream of conventional sand filters. An example is the Gary, Indiana plant, equipped for fully automatic operation. This plant takes water from Lake Michigan as does the installation at Kenosha, Wisconsin. Operation has shown elimination of taste and odor problems due to algae, reduction of chemical dosage, and improved efficiency of the sand filters.

TABLE 3-9 MICROSTRAINING INSTALLATIONS IN OPERATION OR UNDER CONSTRUCTION AND THEIR APPLICATIONS IN NORTH AMERICA AS OF SEPTEMBER, 1969 (COURTESY OF COCHRANE DIVISION-CRANE CO.)

PUBLIC WATER SUPPLIES	Machines	Flow mgd
Primary Treatment Ahead of Rapid Sand Filtration		
1. Baker Metropolitan Water and Sanitation Districts, Colorado	2 — 7½'	6.0
2. Belleville, Ontario	4 — 7½'	9.6
3. Broadmoor Hotel, Colorado Springs, Colo.	2 — 7½'	3.0
4. Estes Park, Colorado	1 — 7½'	2.4
5. Fort Collins-Loveland Water District, Colorado	1 — 7½'	3.5
6. Gary-Hobart Water Corporation, Indiana	1 — 10'	5.0
7. Golden, Colorado	1 — 7½'	3.0
8. Cite des Jeunes, Quebec	1 — 5×3	1.25
9. Kenosha, Wisconsin	4 — 10'	15–25
10. Leamington, Ontario	1 — 10'	4½–9
11. Left Hand Water Supply Company, Longmont, Colorado	1 — 5×3	0.75
12. Little Thompson Water District, Berthoud Colo.	1 — 5×3	1.2
13. Loveland, Colorado — two plants	2 — 10'	10–15
14. Montreal, Quebec — pilot plant	1 — 10'	4–9
15. Public Service Company of Colorado Evergreen — two plants	2 — 5×3	1.25
16. San Gabriel Valley Water Co., Spring St., Fontana, California	1 — 10'	8.0
17. Fairmont, Minnesota	1 — 10'	5.0
18. Greeley, Colorado	2 — 10'	14.0
19. Burlington, Vermont	1 — 10'	8.0
20. Westminster, Colorado	1 — 10'	6.0
Primary Treatment Ahead of Slow Sand Filtration		
1. Ilion, New York	1 — 7½'	2.4
2. Peekskill, New York	2 — 7½	5.0
3. Calgary Power Co. Camrose, Alberta	1 — 7½'	1.5
4. Geneva, New York	2 — 7½'	6.0
Primary Treatment Ahead of Diatomite Filters		
1. Chaumont, New York	1 — 5×1	0.2
2. Edgewood Water Co., Stateline, Nevada	1 — 5×3	1.0
3. Peekskill, New York	1 — 7½'	2.4
Sole Filtration		
1. Bertie Township, Ontario	1 — 10'	5.4
2. Boulder, Colorado	2 — 10'	10–14
3. Brockville, Ontario	3 — 7½'	7.2
4. Calgary Power Co. Camrose, Alberta	1 — 5×3	0.8
5. Coaldale, Alberta	1 — 5×3	0.577
6. Colorado Springs, Colo.	2 — 10'	10.0
7. Covington, Virginia	1 — 7½'	2.4
8. Deaver, Wyoming	1 — 2½'	0.25
9. Denver Board of Water Commissioners, Colo.		
a) Marston Lake Supply 1st Section	12 — 10'	60.0
b) South Side Plant — Pilot Plant	1 — 7½'	2.0

TABLE 3-9 (Continued)

PUBLIC WATER SUPPLIES	Machines	Flow mgd
10. Dunnville, Ontario	6 — 10'	25.6
11. Elk Island National Park, Alberta	1 — 5×1	0.14
12. Fairmont, Minnesota	1 — 10'	4.5
13. Farmington, Maine	1 — 5×3	1.0
14. Fort St. John, British Columbia	1 — 5×3	0.8
15. Geneseo, New York	1 — 7½'	2.4
16. Gleichen, Alberta	1 — 5×3	0.6
17. Glenwood Springs, Colo.	2 — 7½'	4.5
18. Grand Junction, Colo.	3 — 7½'	6.4
19. Harrow, Ontario	1 — 7½'	1.2
20. Hastings, Ontario	1 — 5×3	0.36
21. Honey Brook, Penna.	1 — 2½'	0.15
22. Inuvik, N.W.T.	1 — 5×3	0.4
23. Kill Devil Hills, N.C.	1 — 5×3	0.7
24. Los Alamos, New Mex.	1 — 5×3	0.25
25. Los Angeles, Calif. — Pilot Plant	1 — 2½'	
26. Mistassini, Quebec	1 — 5×3	0.6
27. Montmorency, Quebec	2 — 7½'	4.0
28. Nags Head, N.C.	1 — 7½'	1.5
29. Palisade, Colorado	1 — 7½'	2.0
30. San Gabriel Valley Water Co., Fontana, Calif.		
a) Cherry & San Bernardino Sts.	1 — 7½'	4.0
b) Citrus & Walnut Sts.	1 — 7½'	6.0
31. South River Sanitary Dist., Sharando, Va.	1 — 7½'	2.0
32. Stamford Water Co., Connecticut		
a) North Reservoir	3 — 10'	25.0
b) Waste Water	1 — 5×3	0.75
33. Terrebonne, Quebec	1 — 7½'	1.0
34. Vauxhall, Alberta	1 — 5×3	0.35
35. Vulcan, Alberta	1 — 5×3	0.35
36. Wendell, North Carolina	1 — 5×3	0.75
37. Williams Lake, B.C. — Two plants	2 — 5×3	2.4
38. Palmer Lake, Colorado	1 — 5×3	0.5
39. New York City	2 — 10'	12–16
40. North Kamloops, B.C., Canada	1 — 10'	5.0
41. Sherbrooke, Quebec, Canada	3 — 10'	14.0
42. Lac Etchemin, Quebec, Canada	1 — 5×3	0.6
43. Ohio Water Services, Struthers, Ohio	2 — 10'	13.0
44. Lake Park, Iowa	1 — 5×3	1.0

At Denver, Colorado, an installation of 12–10 ft diameter micro-strainers, treats flows up to 100 mgd. Microstraining is the only filtration employed, the water being taken from a large upland reservoir where turbidity is low but plankton growths occur during spring and summer. Pilot plant tests made when this installation was being designed showed that the microstrainers removed more of the plankton organisms than sand filters.

A microstraining installation is operating in New York City, at outlet from the Jerome Park Reservoir. Two 10 ft diameter microstrainers are used to treat flows up to 14 mgd.

Microstrainers are a tool to solve specialized problems, principally to remove troublesome algae ahead of sand filters. Since it is a surface filter or mechanical strainer, it is subject to many of the same limitations as are diatomite filters in its applications to purification of water from contaminated sources for potable use. Table 3-9 summarizes several different types of installations.

Upflow Filters

Upflow filtration has an obvious theoretical advantage, because coarse-to-fine filtration can be achieved with a single medium such as sand with almost perfect gradation of both pore space and grain size from coarse-to-fine in the direction of filtration (upward). Since the bed is backwashed in the same direction but at higher flow rates, the desired relative positions of fine media are maintained or reestablished with each backwash. The inherent advantage of upflow filtration has long been recognized, and, under laboratory conditions, the anticipated high filtration efficiency has been well-verified by several workers. The difficulty with upflow filtration comes when the headloss above a given level exceeds the weight of the bed above that level, at which time the bed lifts or partially fluidizes allowing previously removed solids to escape in the effluent. In Russia, bed depths up to 6 ft are used in an attempt to minimize bed lifting. In the U. S., parallel plates or a metal grid is placed at the top of the fine media. The spacing of the plates or the size of the openings in the grid are such that the media grains arch across the open space to restrain the bed against expansion. These restraining bar systems have about 75 percent open area in the best designs developed to date. Figure 3-12 illustrates an upflow filter with a restraining grid system. Even with use of a restraining grid or a deep bed, there may be problems with excessive pressures or sudden variations in pressure which break the sand bridge or cause the bed to expand and lose its filter action. The frequency of breakthrough is rare, but the fact that it can occur at all, say with poor operation, has been sufficient to raise questions concerning public health implications and to limit the use of upflow filters for potable water applications. In other areas which are free of health considerations, upflow filters have found wide application and have given excellent service. These areas include process water, wastewater treatment, deep well injection water, API separator effluent, cooling water, and

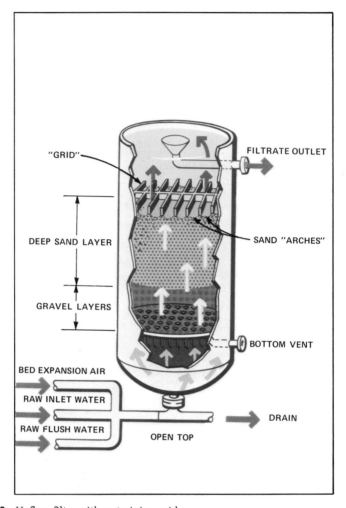

Fig. 3-12 Upflow filter with restraining grid.

other similar applications. Until a foolproof means of restraining a dirty upflow bed is developed, it appears likely that potable water applications of the upflow filter in this country will continue to be limited.

Biflow Filters. Biflow filters are an outgrowth of upflow filters, in that the divided flow (downflow from the top and upflow from the bottom (see Figure 3-13) is an attempt to restrain the bottom upflow portion of the bed by placing above it a downflow filter. Biflow filters are used in Holland and Russia but not to any extent in the United States. Biflow

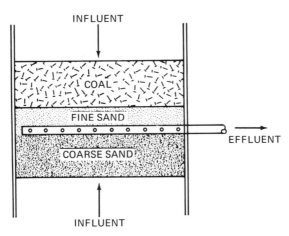

Fig. 3-13 Section through a dual media biflow filter.

filters permit filtration in two opposite directions at the same time. Essentially the top and bottom halves are completely independent filters equal in capacity which results in some savings in structure and underdrains.

Unfortunately, the biflow filter has an inherent limitation which seems to preclude development of a unit which will produce an exceptionally high quality effluent. First consider a single media biflow bed. The finest materials are at the top of the upper downflow bed. This makes the top half of the bed a rapid sand or surface type filter and the quality of water produced at best cannot exceed that from a rapid sand filter. The bottom half of this same filter is a coarse-to-fine filter, but unfortunately the finest material at the top outlet from the bed is somewhat coarser than the finest material which can be successfully used in a rapid sand filter. Obviously, the effluent from this bed will be of lesser quality than that from the rapid sand downflow filter above. This fact has been recognized by others and revealed by pilot tests. This lead to a consideration of the dual media biflow filter as illustrated in Figure 3-13. The idea here and the advantage over the single media biflow bed is that this arrangement places the fine sand closer to the mid-collector. It provides a dual media (coal-sand) downflow bed above a coarse-to-fine single media (sand) upflow bed. Again the limitation is the coarseness of the finest sand which must be used from a practical standpoint. If the sand is finer than that ordinarily used in a rapid sand filter as it must be to build the best upflow filter in the bottom half of the bed, then the sand will be so fine that excessive amounts of sand will rise into the coal bed during backwashing. The gradation of a sand which will be suitable for the dual media bed in the upper half of the bed will be too coarse to provide the best possible filtration in the upflow bottom half of the bed. The quality of effluent from either half will not approach that from a mixed-media filter. This problem is so basic that there does not appear to be an easy solution to this dilemma, but perhaps one will be found.

Deep Coarse Beds.   Deep coarse single media filters are finding increasing acceptance in wastewater treatment, particularly attached growth nitrification and denitrification systems. While they have not yet found widespread application in potable water production, they are of interest because of their future possibilities.

Deep bed filtration systems use media from 1 to 6 mm in diameter in beds 6 to 10 ft deep and operate at rates of 1.5 to 15 gpm/sf. The media is silica sand chosen for its sphericity and is closely sized to give as low a uniformity coefficient as possible. Air is used for backwashing and water is applied simultaneously to flush accumulated suspended material from the unit. Filter bottoms are nozzleless concrete blocks. Loadings obtained with these deep bed systems range from 1 to 10 lb/cf/cycle. In-depth filtration is obtained because of the large size of the media particles used even with application of heavy loadings of relatively coarse solids. Only 10 percent bed expansion is obtained in backwash which avoids classification of the media during this operation.

There are many operating installations of deep coarse bed filters principally on industrial process water and wastewater, but there may be principles and experience here which can be transferred with benefit to potable water production.

Moving Bed Filter.   The old "drifting sand filter" is being rejuvenated with modern mechanization to handle exceptionally heavy loads of applied solids in the form of the MBF (moving bed filter). Because of the heavy involvement of equipment in the process, the maximum unit size is presently limited to about ¼ mgd. These characteristics make the MBF more useful for relatively small wastewater treatment applications than for potable water production, but, again, the principles are of interest here. The MBF is a continuous sand filter in which the water moves countercurrent to the sand. The sand is driven through a cone in one direction while simultaneously passing the water to be treated through the filter bed in the opposite direction. The solids at the inlet end of the filter are removed from the bed along with the dirty sand as rapidly as required by their buildup. Movement of the sand is accomplished by means of a hydraulically actuated diaphragm. The diaphragm pushes the sand bed as a plug through the cone toward the inlet end. Clean sand feeds by gravity into the void left in the bed in front of the diaphragm. The pushing cycle is then repeated. The sludge sand mixture is passed through a washing tower and the clean sand is returned to the feed hopper of the MBF. Because the unit is continuously cleaned, it is possible to use sand finer than that contained in a conventional rapid sand filter, so that some improvement in effluent quality might be obtained.

REFERENCES

1. American Water Works Association, "State of the Art of Water Filtration," Committee Report, p. 662 (1972).
2. American Water Works Association, "Diatomite Filters For Municipal Use," Task Group Report, *Journal American Water Works Association*, p. 157 (1965).
3. Baylis, John R., "Progress in the Design of Rapid Sand Filters," Transactions 2nd Annual Conf. on San. Eng., Univ. of Kansas, p. 7 (1952).
4. Baylis, John R., "Seven Years of High Rate Filtration," *Journal American Water Works Association*, p. 585 (1956).
5. Bernade, M. A., and Johnson, Barry, "Schistosome Cercariae Removal by Sand Filtration," *Journal American Water Works Association*, p. 449 (1971).
6. Calise, V. J., and Homer, W. A., "Russian Water Supply and Treatment Practices," *J. San. Eng. Div., Proceedings American Society of Civil Engineers*, p. 1 (March 1960).
7. Camp, T. R., "Theory of Water Filtration," *Proceedings American Society of Civil Engineers*, p. 1 (1964).
8. Cleasby, J. L., "Filter Rate Control Without Rate Controllers," *Journal American Water Works Association*, p. 181 (1969).
9. Cleasby, J. L., "Approaches to a Filtrability Index for Granular Filters," *Journal American Water Works Association*, p. 372 (1969).
10. Cleasby, J. L., Williamson, M. M. and Baumann, R. E., "Effect of Filtration Rate Changes on Quality," *Journal American Water Works Association*, p. 869 (1963).
11. Conley, W. R. Jr., and Hsiung, K., "Design and Application of Multimedia Filters," *Journal American Water Works Association*, p. 97 (1969).
12. Conley, W. R. Jr., "Experiences with Anthracite Sand Filters," *Journal American Water Works Association*, p. 1473 (1961).
13. Conley, W. R. Jr., and Pitman, R. W., "Innovations in Water Clarification," *Journal American Water Works Association*, p. 1319 (1960).
14. Conley, W. R. Jr., "Integration of the Clarification Process," *Journal American Water Works Association*, p. 1333 (1965).
15. Craft, T. F., "Comparison of Sand and Anthracite for Rapid Filtration," *Journal American Water Works Association*, p. 10 (1971).
16. Culbreath, Mark C., "Necessary Modifications for High Rate Filtration," *Journal American Water Works Association*, p. 699 (1968).
17. Culp, G. L., "Secondary Plant Effluent Polishing," *Water & Sewage Works*, p. 145 (1968).

18. Culp R. L., and Culp, G. L., *Advanced Wastewater Treatment*, Van Nostrand Reinhold Co., New York, 1971.
19. Culp, R. L., "Filtration," Chapter 8, *Water Treatment Plant Design Manual*, American Water Works Association, New York, 1969.
20. Culp, R. L., "New Water Treatment Methods Serve Richland," *Public Works*, p. 86 (July 1964).
21. Eunpu, Floyd F., "High Rate Filtration in Fairfax County Virginia," *Journal American Water Works Association*, p. 340 (1970).
22. Fair, G. M. and Geyer, J. C., *Water Supply and Wastewater Disposal*, John Wiley & Sons, New York, 1954.
23. Frissora, F. V., "An Advanced Water Filtration Plant," *Water & Sewage Works*, p. 365 (1971).
24. Geise, G. D., Pitman, R. W., and Wells, G. W., "The Use of Filter Conditioners in Water Treatment," *Journal American Water Works Association*, p. 1303 (1967).
25. Hamann, Carl L. and McKinney, Ross E., "Upflow Filtration Process," *Journal American Water Works Association*, p. 1023 (1968).
26. Diaper, E. W. J. and Ives, K. J., "Filtration Through Size Graded Media," *J. San. Eng. Div., Proceedings American Society of Civil Engineers*, p. 89 (June 1963).
27. Harmeson, R. H. *et al.*, "Coarse Media Filtration for Artificial Recharge," *Journal American Water Works Association*, p. 1396 (1968).
28. Harris, W. Leslie, "High-Rate Filter Efficiency," *Journal American Water Works Association*, p. 515 (1970).
29. Hudson, H. E. Jr., "High Quality Water Production and Viral Disease," *Journal American Water Works Association*, p. 1265 (1962).
30. Hudson, H. E. Jr., "Physical Aspects of Filtration," *Journal American Water Works Association*, p. 33 (1969).
31. Hudson, H. E. Jr., "Theory of the Functioning of Filters," *Journal American Water Works Association*, p. 868 (1948).
32. Hudson, H. E. Jr., "Functional Design of Rapid Sand Filters," *J. San. Eng. Div., ASCE*, p. 17 (1963).
33. Ives, K. J., "Progress in Filtration," *Journal American Water Works Association*, p. 1225 (1964).
34. Kreissl, J. F., Robeck G. G., and Sommerville, G. A., "Use of Pilot Filters to Predict Optimum Chemical Feeds," *Journal American Water Works Association*, p. 299 (1968).
35. Laughlin, J. E., and Duvall, T. E., "Simultaneous Plant Scale Tests of Mixed Media and Rapid Sand Filters," *Journal American Water Works Association*, p. 1015 (1968).
36. Minz, D. M., "Modern Theory of Filtration," Special Subject No. 10, International Water Supply Assoc., London, 1966.
37. Moffett, J. W., "The Chemistry of High Rate Water Treatment," *Journal American Water Works Association*, p. 1255 (1968).
38. Mohanka, S. S., "Multilayer Filtration," *Journal American Water Works Association*, p. 504 (1969).
39. Moulton, F., Bedc, C., and Guthrie, J. L., "Mixed-Media Filters (Clean) 50 MGD without Clarifiers," *Chemical Processing*, (March 1968).

40. Nelson, O. F., "Capping Sand Filters," *Journal American Water Works Association*, p. 539 (1969).

41. Oakley, H. R., and Cripps, T., "British Practice In the Tertiary Treatment of Sewage," *Journal Water Pollution Control Federation*, p. 36 (1969).

42. Oakley, H. R., personal communication, 1969.

43. Oeben, R. W., Haines, H. P., and Ives, K. J., "Comparison of Normal and Reverse-Graded Filtration," *Journal American Water Works Association*, p. 429 (1968).

44. O'Melia, C. R., and Crapps, D. K., "Some Chemical Aspects of Rapid Sand Filtration," *Journal American Water Works Association*, p. 1326 (1964).

45. O'Melia, C. R. and Stumm, Werner, "Theory of Water Filtration," *Journal American Water Works Association*, p. 1393 (1967).

46. Pitman, R. W. and Wells, G. W., "Activated Silica as a Filter Conditioner," *Journal American Water Works Association*, p. 1167 (1968).

47. Rice, A. H., and Conley, W. R., "The Microfloc Process in Water Treatment," *Tappi*, p. 167A (1961).

48. Riddick, T. M., "Filter Sand," *Journal American Water Works Association*, p. 121 (1940).

49. Robeck, G. G., Clarke, N. A. and Dostal, K. A., "Effectiveness of Water Treatment Processes in Virus Removal," *Journal American Water Works Association*, p. 1275 (1962).

50. Robeck, G. G., Dostal, K. A., and Woodward, R. L., "Studies of Modifications in Water Filtration," *Journal American Water Works Association*, p. 198 (1964).

51. Sedore, J. K., "We Converted to High Rate Filtration," *Amer. City*, April, 1968.

52. Segall, B. A. and Okun, D. A.,"Effect of Filtration Rate on Filtrate Quality," *Journal American Water Works Association*, p. 468 (1966).

53. Seth, A. K. et al., "Nematode Removal by Rapid Sand Filtration," *Journal American Water Works Association*, p. 962 (1968).

54. Shea, T. G., Gates, W. E., and Argaman, Y. A., "Experimental Evaluation of Operating Variables in Contact Flocculation," *Journal American Water Works Association*, p. 41 (1971).

55. Shull, K. E., "Experiences with Multi-bed Filters," *Journal American Water Works Association*, p. 314 (1965).

56. Smit, P., "Upflow Filter," *Journal American Water Works Association*, p. 804 (1963).

57. Smith, C. V. Jr., and Medlar, S. J., "Filtration Optimization Utilizing Polyphosphates," *Journal American Water Works Association*, p. 921 (1968).

58. Syrotynski, S., "Microscopic Water Quality and Filtration Efficiency," *Journal American Water Works Association*, p. 237 (1971).

59. Tuepker, J. L., and Buescher, C. A. Jr., "Operation and Maintenance of Rapid Sand and Mixed-Media Filters in a Lime Softening Plant," *Journal American Water Works Association*, p. 1377 (1968).

60. Walker, J. D., "High Energy Flocculation and Air-Water Backwashing," *Journal American Water Works Association*, p. 321 (1968).

61. Walton, Graham, "Effectiveness of Water Treatment Processes," U. S. Public Health Service, Publication No. 898, 1962.

62. Westerhoff, G. P., "Experience With Higher Filtration Rates," *Journal American Water Works Association*, p. 376 (1971).
63. Willis, J. F., "Direct Filtration, an Economic Answer to Water Treatment Needs," *Public Works*, p. 87 (Nov. 1972).
64. Woodward, R. L., "Relation of Raw Water Quality to Treatment Plant Design," *Journal American Water Works Association*, p. 432 (1969).
65. Yao, K. M., Hubibjun, M. T., and O'Melia, C. R., "Water & Waste Filtration-Concepts and Application," *ES&T*, p. 1105 (Nov. 1971).

4 Coagulation Control and Monitoring

TECHNIQUES AVAILABLE

All of the excellence provided in physical treatment plant design and equipment selection is worthless unless the chemical coagulation of the raw water being treated is properly carried out. Inadequate doses of coagulant will result in a finished water of excessive turbidity. Excessive doses may also cause this result and are wasteful of the public's funds. Excessive turbidity makes the water aesthetically unacceptable as well as increasing the probability of microorganisms escaping subsequent disinfection processes. Public acceptance of the entire water supply program is dependent upon achieving proper coagulation.

Since achieving proper coagulation has been a problem universally faced by water treatment operators for many years, a wide variety of techniques have been developed for controlling the coagulation process. Most of these involve laboratory tests, the results of which then must be manually transferred to the full-scale plant operation by the plant operator. For example, the most widely used technique has been the conventional jar test. A sample of raw water is collected and transported to the laboratory where it is split into several samples in beakers. Varying coagulant doses are added to each sample and, after an appropriate period of mixing by a multiple paddle mixing device, stir-

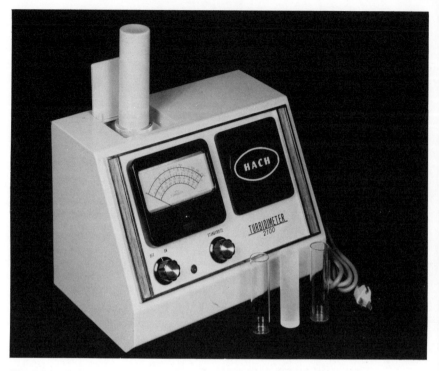

Fig. 4-1   Laboratory turbidimeter. *(Courtesy Hach Chemical Co.)*

ring is stopped and the floc is allowed to settle. The clarity of the supernatant is then used to determine the optimum coagulant dose. In many cases, the operator's visual inspection of the supernatant is all that is used to select the best dose. Measurement of filtered supernatant turbidity in a batch, laboratory turbidimeter (see Figure 4-1) provides a more objective means of selecting the best dose of those tested. Whether or not the coagulant dose which provides the optimum result in a beaker will also provide optimum results on a plant scale depends on the unlikely fact that the same efficiency of mixing is achieved in both cases. Also, the fact that the procedure is a batch procedure results in an inherent time lag in responding to changes in raw water conditions. This lag may be only an hour if the operator is on duty and alert to raw water conditions but it may be several hours if he is off duty or involved in another task, such as equipment maintenance. This method may give satisfactory — even if not optimum — results when applied to a raw water of relatively uniform quality and which contains only a moderate amount of organic turbidity, For low turbidity waters and waters containing large amounts of organic material, no sharp distinction between coagulant doses near optimum may exist with the result that it is difficult to determine from the appearance of the sample whether adequate plant-scale results will be obtained. Although adequate control can be achieved under certain conditions by visual monitoring of jar tests, at other times it becomes a guessing game with each observer seeing characteristics which may or may not affect plant performance.

A recent review of available coagulation control techniques[1] lists 22 alternate techniques which fall into three general categories: (a) conventional and modified jar tests where visual observations of supernatant, floc formation time, floc density, etc., are used, (b) techniques based on particle charge, and (c) techniques based upon filtering the coagulated water and measuring the filtrate turbidity. As the jar test and its many variations have been discussed in many articles and books as well as being used for many years, it will be discussed here only as applied to evaluation of potential coagulant aids.

JAR TESTS

A standard jar test procedure is contained within AWWA manual M-12 titled *Simplified Procedures for Water Examination*. A standard testing plan is required to provide a basis for the comparison of polymers as primary coagulants and for coagulation aids in a centrally directed program.

Manual M-12 states "the jar tests are designed to show the nature and extent of the chemical treatment which will prove effective in the full scale plant. Many of the chemicals added to a water supply can be evaluated on a laboratory scale by means of jar tests. Among the most important of these chemicals are coagulants, coagulant aids, alkaline compounds, softening chemicals, and activated carbon for taste and odor removal. The test also permits the appraisal of the relative merits of the aluminum and iron coagulants, or in conjunction with such coagulant aids as activated silica, polyelectrolytes, clays, stone dust, activated carbon, settled sludge, lime, and soda ash."

The manual also notes several considerations that may affect the outcome of jar testing. It states "even the smallest detail may have an important influence on the result of a jar test. Therefore, all samples in a series of tests should be handled as nearly alike as possible. The purpose of the test will determine such experimental conditions such as flocculation and settling intervals. Since temperature plays an important role in coagulation, the raw water samples should be collected and measured only after all other preparations have been made, in order to reduce the effect that room temperature might have on the sample."

The following equipment may be utilized to evaluate polymers as coagulant aids:[11]

1. A variable speed (0 to 100 rpm) stirring machine with 3 to 6 paddles (each paddle size 1 to 3 in.).
2. A floc illumination device to be located under the base of the laboratory stirrer. This illuminator will be used to determine both the floc formation time and the visual floc size included on the data sheet.
3. 6 standard beakers of 1500 ml capacity (approximately 4.5 in. diameter by 6.4 in. high). These pyrex beakers will be required as reaction vessels.
4. A plastic pail or similar sample collection container with a capacity in excess of 2 gallons will be required. The sample should be homogeneous for each complete jar test run.
5. One 1000 ml graduated cylinder.
6. Measuring pipets of 1, 5, and 10 ml, graduated in 0.1 ml steps, for dosing samples rapidly with the necessary coagulants, suspensions, and solutions. These pipets should be thoroughly rinsed with distilled water after using to prevent the caking of solutions.
7. A 100 ml pipet for withdrawing the coagulated softened water supernatant sample for turbidity and hardness determinations.
8. Instrumentation for the monitoring of water temperature, turbidity, floc formation time, and hardness (for softening plants) is required.
9. Special apparatus for softening test:
   a. Filter Funnel
   b. Filter paper of medium retentiveness. Either Whatman No. 40 or Schleicher & Shull No. 589 is satisfactory.

The importance of simulating plant conditions as closely as possible cannot be overemphasized. The manual states "dosing solutions or suspensions should be prepared from the stock materials actually used in plant treatment. Distilled water used for the preparation of lime suspensions should be boiled for 15 minutes to expel the carbon dioxide and then cooled to room temperature before the lime is added." Details are presented in AWWA Manual M-12 on the preparation of coagulant dosing solutions and suspensions.

In testing polymeric coagulants or coagulant aids supplied by cooperating polymer manufacturers, solutions of these products should be prepared in accordance with manufacturer specifications.

Jar Test Procedure for Coagulant Aid Treatment

The addition of alkaline agents, clay suspensions, or activated silica are covered under number 14 in the following format as presented in reference 11. When the water contains natural color, alum should be added before other agents or coagulant aids.

1. Rinse six 1500 ml beakers with tap water and let the beakers drain for a few minutes in an upside down position. Beakers that have had several days of use should be scrubbed inside and out with a brush and a household dishwashing detergent, finishing with a thorough rinse of tap water.

2. Clean the stirring machine paddles with a damp cloth.

3. Collect a sample of raw water and complete steps 4 through 8 within 20 min. Throw away the sample and collect a fresh sample if the work must be interrupted during this critical stage. Otherwise, the settling of high turbidity and an increase in the sample temperature from the heat of the laboratory may cause erroneous results.

4. Stand the beakers right side up and pour more than 1 liter of raw water into each one.

5. Taking one beaker at a time, pour some of the raw water back and fourth between the beaker and a 1 liter graduated cylinder. Finally, fill the graduated cylinder to the 1 liter mark, and discard the excess raw water in the beaker. Return the measured 1 liter sample to the beaker.

6. Place all the beakers containing the measured 1 liter samples on the stirring machine.

7. Lower the stirring paddles into the beakers, start the stirring machine and operate it at 100 rpm.

8. With a measuring pipet, add the same dosages of the coagulant solution that is used in your treatment facility to the 6 test beakers, at nearly the same time as possible. Record the time required to produce a visible floc on the data sheet.

9. Apply the polymeric coagulant aid at dosages of 0.05, 0.1, 0.15, 0.2 ppm, etc. to the 5 above beakers at 100 rpm for 3 min. The 6th beaker should be a control jar to which no coagulant aid is to be added. With a proper series, the beakers should show good to excellent coagulation results depending on the control jar. It may be necessary to repeat the jar test to ascertain the proper series of doses for the desired results.

10. Reduce the stirring speed over the next 30 sec to 40 rpm, and continue stirring at this speed for a 15 min period.

11. Stop the stirring machine and allow the samples to settle for 5 min. Observe the floc sizes in each jar. Use the control jar floc size as unity. Then visually compare the size of the remaining floc particles in jars 1 to 5 under visual floc size, an example, if the floc in jar 5 appears twice as large as the floc in control jar 6, record 2 for jar 5 under visual floc size in the data sheet.

12. Using a 100 ml pipet, withdraw a sample for turbidity measurement 1 in. below the surface of each beaker.

13a. Determine the turbidity of the coagulated sample in accordance with *Standard Methods* or approved turbidimeters. Record this on the data sheet.

  b. Using the sample from the turbidity measurement, determine the supernatant pH and record the pH on the data sheet.

14a. Add alkaline agents if the plant normally adds these agents.

  b. Add clay suspensions. Where clay is used to improve the coagulation of low turbidity water, add a well-shaken clay, kaolin, or stone dust dosing suspension (prepared in accordance with dosing solution and suspensions specifications) just before the coagulant. Then follow procedures 1 to 13.

  c. Addition of activated silica. Where activated silica is employed as a coagulant aid, use the activated silica sol in amounts to give a dosage range of 1 to 7 mg/l. On the average, activated silica dosages of 2 to 5 mg/l yield satisfactory results. Determine by experiment whether the activated silica sol should be added before or after the coagulant for best results. Then follow the procedure in items 1 to 13.

TECHNIQUES BASED ON PARTICLE CHARGE

Zeta Potential

The first technique of this type to receive attention as a possible means of coagulation control was the zeta potential technique. This technique has received a great deal of published attention and the few references [3-6] shown at the end of this chapter will provide the reader with a more extensive bibliography as well as a more detailed introduction to the subject if he so desires.

Zeta potential is a measure, in millivolts, of the electrical potential existing between the outer and fixed layer of counterions surrounding colloids and the bulk of the liquid in which the colloid is suspended. Because like charges repel, colloids in water — which are almost always negatively charged — resist coagulation. When this negative zeta potential is reduced, the repulsive forces are likewise reduced, and if agitated gently, the colloids will flocculate. In the treatment process, the reduction of zeta potential is accomplished by the addition of a positively charged ion or complex from such coagulants as aluminum sulfate, the iron salts, and cationic polyelectrolytes. The zeta potential is determined by observing microscopically visible particles as they move in an electric field. The particle zeta potential may be calculated from the equation:

$$\zeta = \frac{4\pi\eta V}{HD}$$

in which ζ, is the zeta potential in volts; η, is the viscosity in poises; V, the particle velocity in centimeters per second; H, the potential gradient in volts per centimeter; and D, the dielectric constant.

Since zeta potential is directly proportional to mobility, and small undeterminable changes will occur in the viscosity and the dielectric constant, most workers in this field express results in terms of particle mobilities.

The mobility may be calculated from the equation:

$$\Omega = \frac{dx}{tIR_s}$$

in which d is the distance across a square of the Howard counter at a given magnification; x, the cross sectional area of the cell in square centimeters; t, the time in second; I, the current in amperes, and R_s, the specific resistance of the suspension in ohm-centimeters. The experimental results are thus expressed in velocity per unit field strength — that is, in μ per sec per volt per cm.

Mobility values are expressed as positive (+) for those particles which migrate toward the negative pole of the cell, and as negative (−) for those that migrate toward the positive pole. The point at which there is no particular migration (zero mobility) is considered to be the isoelectric point.

The use of this technique is also a batch procedure, as is the jar test. The general procedure involves varying the coagulant dosage and measuring the resulting zeta potential. However, each water requires comparative tests to determine the correlation between zeta potential and finished water turbidity in the plant. Organic colloids such as those constituting organic color generally require zeta potentials near zero while clay-related turbidity is best removed at somewhat negative zeta potentials. Typical values for optimum coagulation range from +5 to −10 millivolts, depending upon the nature of the material to be removed.

The procedure for measuring zeta potential is somewhat tedious to expect that it would be used routinely in the average water treatment plant. Black[3] presented the following procedure for using the Briggs cell in conjunction with a microscope for zeta potential measurement:

1. The clean cell is fitted with the electrode vessels.
2. The sample (in electrolyte medium) whose mobility is to be determined is placed in the cell. Suction lines from a water aspirator

are applied to the ends of the cell to draw the sample in both directions from the funnel until all air bubbles are removed. The dark field condenser lens is covered with immersion oil and lowered below the level of the stage. The cell, in the holder designed to keep it level, is placed on the microscope stage and the electrode leads are attached to the platinum wires extending from the bottom of the electrode vessels.

3. The dark field condenser is raised until a round bubble appears on the bottom of the cell. The cell should be flat on the microscope stage and should not be moved by raising the condenser too far.

4. The light source and microscope mirror are adjusted for proper illumination. The light beam must pass through the $CuSO_4$ solution to absorb heat and to prevent convection currents in the cell.

5. The microscope is focused on the bottom of the cell using the proper illumination.

6. The fine adjustment on the microscope is then rotated the calculated number of divisions necessary to come up to the bottom stationary layer.

7. The switch on the control box is turned on and, using the potentiometer, the resistance is adjusted until the speed of the particles can be easily timed. (Analyst should avoid touching both electrodes simultaneously when the current is on, as sufficient voltage is present to give a shock.)

8. The milliammeter reading is recorded and kept constant throughout each individual sample.

9. With a stop watch, particles are timed as they travel a known distance on the cross hatch. The movement should be in a horizontal direction only, from one electrode to another.

10. Approximately 10 particles should be timed going in each direction and the direction of movement after each observation should be altered by using the polarizing switch. There should be agreement in these readings in each direction if the stationary layer is properly obtained and there are no leaks in the system.

11. If the electrodes gas during a determination, or if there appears to be a leak in the system, particles move when the current is off, the cell should be dismantled, cleaned, and new electrodes employed in subsequent determinations.

12. As the field strength (volts per centimeter) within the cell is determined by use of Ohm's law, resistance values of suspensions are needed. These are determined with a conductivity bridge using a calibrated conductivity cell. The conductivity must be measured on part of the suspension which has not been in the cell.

13. As pH is closely related to particle mobility, the pH of each sample should be taken.

A typical calculation of mobility, is as follows:

given:

$d = 49\ \mu$;

$x = 1.73$ cm $\times$ 0.100 cm = 0.173 cm^2;

$t = 7.8$ sec (average);

$I = 4.0 \times 10^{-4}$ amp;

$R = 4450$ ohms; $R_s = 4450/1.0 = 4450$ ohm-cm;

$$\Omega = \frac{(49)\ (0.173)}{(7.8)\ (4.0 \times 10^{-4})\ 4450}$$

$$= 0.61\ (\mu/\text{sec}/V/\text{cm})$$

Although zeta potential measurements are useful as a research tool, it is generally agreed that this method is not easily adapted to the typical treatment plant because of the considerable degree of skill and patience required to make the measurements and the amount of interpretation required to make the data useful. It has been the authors' experience in visiting plants which have purchased zeta meter equipment for routine operation that generally it will be found stored and inoperative. Other observers[1] also report an opinion that the "popularity is declining recently" for zeta potential as a control technique. It is also subject to the same shortcomings of any other batch test in that sudden changes in raw water conditions may not be detected until a poor quality finished water occurs in the full-scale plant.

Streaming Current Detector

A streaming current detector may be used to provide a continuous measure of the relative charges. This technique involves placing a sample in a special cylinder containing electronic sensing electrodes at the top and bottom. A loose-fitting piston is then partly submerged into the sample, and is reciprocated along its axis to produce an alternating current between the electrodes when the cylinder contains moving charges. A synchronous motor drives the piston and a synchronous rectifier switch by means of which the alternating current generated by the alternating fluid motion is made to register on a dc meter (Figure 4-2). An amplifier with adjustable negative feedback is used to provide an output proportional to the current collected by the electrodes. Readout may be by a microammeter, with calibration in arbitrary units. The alternating current is analyzed and related to zeta potential, which can be used to control coagulant dosage. A flow-through cylinder and recording ammeter may be built into the system to permit continuous monitoring. Almost all of the successful

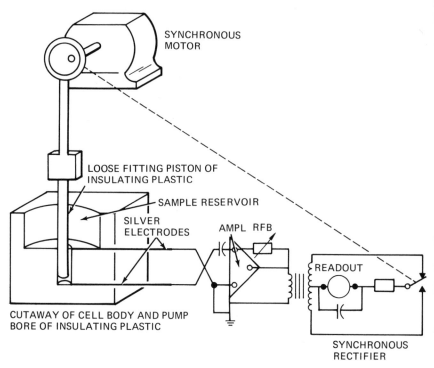

Fig. 4-2 Simplified diagram of SCD instrument.

applications of the instrument have involved the titration of charge-influencing materials, either batchwise or continuously. In these applications, the uncertainty as to what is measured is of little consequence so long as the addition of anionic and cationic materials changes the reading in a reproducible manner. Titrations are carried out in the same fashion as acid-base titrations with a pH meter. Continuous readings are obtained while adding one material to the other.

The use of this approach for continuous control of coagulation is subject to the other limitations of the zeta potential approach, including definite limitations when used to evaluate synthetic polymer coagulants such as when the polymer is anionic and is used to coagulate negatively charged particles. Because the technique involves only the aspects of electronic charge, it may lead to erroneous conclusions when used to study coagulants not following the electrokinetic theory. As with zeta potential, the data must be correlated with the usual indices of plant performance as there may not be any consistent relationship between charge and filtered water clarity even at a given plant for various seasons of the year. However, the continuous nature of the streaming current detector may make it attractive in some instances where cationic coagulants are used in a water which shows little variation from season to season.

Colloid Titration

A simple titration technique using an indicator that changes colors when a solution is titrated to electric neutrality has been developed.[7,8] An excess amount of positively charged polymer is added to the naturally negatively-charged water. It is then back-titrated with a standard negatively-charged polymer. An empirical correlation is used to relate the volume of titrant to the proper coagulant dosage, in a manner analogous to zeta potential measurement.

Like the streaming current detector, colloid titration has the advantage of using a larger, more representative sample than is used in zeta potential measurements. Since its conclusions are based on the electrokinetic properties of a suspension, its results are subject to the same limitations given for the streaming current detector and the zeta potential devices.

CONTINUOUS FILTRATION TECHNIQUES

Pilot Filters

Because the goal of the water treatment plant should be to produce the minimum possible filter effluent turbidity at, preferably, the minimum chemical costs, it is logical that perhaps the best measure of the efficiency of the coagulation-filtration steps would be the direct, continuous measurement of the turbidity of coagulated water which has passed directly through a pilot granular filter. The application of continuous turbidity monitoring equipment to the effluent of the plant scale filters should be a must for monitoring plant performance but has limited value as a control technique due to the substantial lag time between the point of chemical coagulant addition and the point of filtrate turbidity monitoring. For example, improper coagulation of the incoming raw water will result in a clarifier (with typically 1 1/2–3 hr of detention time) being filled with improperly coagulated water before the results will become fully apparent at the discharge of the plant filters.

The pilot filter technique is applied by sampling the plant treated, coagulated water from the discharge of the plant rapid mix basin. This sample stream is than passed through a small (usually 4 1/2 in. diameter) pilot filter to determine if the coagulant dose is proper by continuously monitoring the pilot filter effluent turbidity. This technique provides a continuous, direct measurement of the turbidity which can be achieved by filtration of the water as it has been coagulated in the actual plant. Thus, no extrapolation from small-scale laboratory coagulation experiments is required. *The only purpose is to determine the proper coagulant dose.* It is not to predict the length of filter run, not to determine the optimum filter aid dose, nor to predict the rate of head-loss buildup. Pilot filter columns have been used for these other purposes but the technique of interest in this chapter is that used for monitoring the coagulation process.

The pilot filter technique has the advantages of offering a continuous monitoring of the plant-scale coagulation process with a minimum lag time. Filtering the water through the pilot filter yields an immediate answer as to the adequacy of the coagulant dosage. In a typical situation, correctness of coagulant dose is determined within 10 to 15 min after the raw water enters the plant. Experience at many locations shows that the pilot filter effluent turbidity is a very accurate predic-

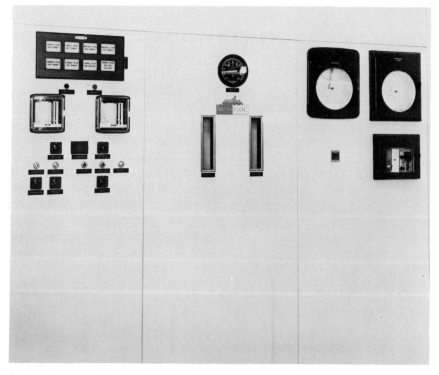

Fig. 4-3 Typical pilot filter installation in plant control panel. (*Courtesy Neptune Microfloc, Inc.*)

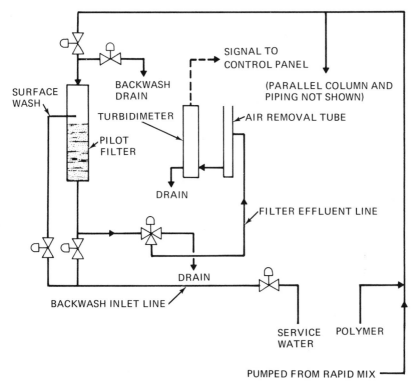

Fig. 4-4 Schematic diagram of a pressure pilot filter system. *(Courtesy Neptune Microfloc, Inc.)*

tion of the turbidity that will be produced by the plant scale filter when the corresponding water reaches the plant scale filter.

The mixed-media filter bed design described in the previous chapter is often used in the pilot filter. The mixed-media design has the ability to accept the high solids load associated with most unsettled, coagulated waters without excessive headloss buildup. The turbidity of the pilot filter is monitored continuously and recorded. High turbidity in the filter effluent could result from either an improper coagulant dose or a breakthrough of floc from a properly coagulated water. To insure that breakthrough does not occur, supplemental doses of polymers are injected into the pilot filter influent line. These large polymer doses may shorten the pilot filter run times but, to insure that breakthrough does not occur, it is desirable to backwash the filter every 1–3 hr in any case to insure that it is always relatively clean. To insure that a continuous monitoring of the coagulation process is provided, 2 pilot filters are used in parallel. The system is equipped so that the filters backwash automatically on a high headloss signal. When one pilot

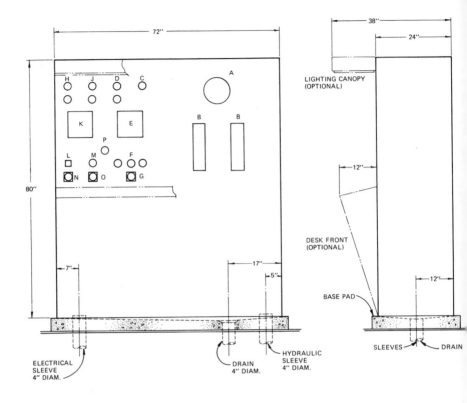

PANEL IDENTIFICATION:
A. CONTROL FILTER HEADLOSS INDICATOR (0-25 FT. WATER)
B. CONTROL FILTER VIEWPORTS
C. CONTROL FILTER HIGH HEADLOSS ALARM LIGHT
D. CONTROL FILTER HIGH & LOW TURBIDITY ALARM LIGHTS
E. CONTROL FILTER TURBIDITY RECORDER (1 PEN)
F. CONTROL FILTER OPERATION INDICATING LIGHTS
G. CONTROL FILTER OPERATION SELECTOR SWITCH
H. PLANT FILTER HIGH & LOW TURBIDITY ALARM LIGHTS
J. CLEARWELL HIGH & LOW TURBIDITY ALARM LIGHTS
K. PLANT FILTER & CLEARWELL TURBIDITY RECORDER (2 PEN)
L. PLANT FILTER SELECTION INDICATING LIGHT
M. CLEARWELL SAMPLE PUMP INDICATING LIGHT
N. PLANT FILTER SELECTION SWITCH
O. CLEARWELL SELECTION SWITCH
P. ALARM SILENCE PUSH BUTTON

Fig. 4-5 Typical console unit housing pilot filter system. (*Courtesy Neptune Microfloc, Inc.*)

filter enters the backwash cycle (which requires only about 10 min), the other pilot filter is automatically placed in service. Where very high raw water turbidities can be expected, a miniature flocculator-tube settler device can be installed ahead of the pilot filters to insure reasonable filter run times even during periods of high turbidity.

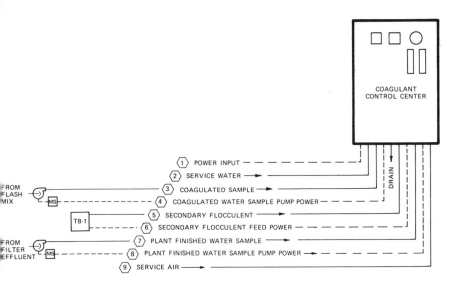

COAGULANT CONTROL CENTER

(1) POWER INPUT — — — — — — — — — —

(2) SERVICE WATER →

FROM FLASH MIX (3) COAGULATED SAMPLE →

MS (4) COAGULATED WATER SAMPLE PUMP POWER — — — — —

TB-1 (5) SECONDARY FLOCCULENT →

(6) SECONDARY FLOCCULENT FEED POWER — — — — — — — — —

FROM FILTER EFFLUENT (7) PLANT FINISHED WATER SAMPLE →

MS (8) PLANT FINISHED WATER SAMPLE PUMP POWER →— — — — — — —

(9) SERVICE AIR →

DRAIN

PIPING AND WIRING REQUIREMENTS:

(1.) POWER INPUT: 120/240 VOLTS, 40 AMPS, THREE WIRE, SINGLE PHASE, 60 CYCLE

(2.) SERVICE WATER: 5/8-INCH ID COPPER TUBING FOR 0-3 GPM AT 20 PSI

(3.) COAGULATED SAMPLE: 5/8-INCH ID COPPER TUBING FOR 1 GPM AT 25 PSI MINIMUM AT INLET TO COAGULANT CONTROL CENTER.

(4.) COAGULATED WATER SAMPLE PUMP POWER: 120 VOLTS, 3 #12 WIRES

(5.) SECONDARY FLOCCULENT FEED: 1/4-INCH POLYETHYLENE TUBING FOR 0-20 GPD

(6.) SECONDARY FLOCCULENT FEED POWER: 120 VOLTS, 2 #12 WIRES

(7.) PLANT FINISHED WATER SAMPLE: 5/8-INCH ID COPPER TUBING FOR 1 GPM AT 25 PSI

(8.) PLANT FINISHED WATER SAMPLE PUMP POWER: 120 VOLTS, 2 #12 WIRES

(9.) SERVICE AIR 80 PSIG

Fig. 4-6 Piping and wiring required for pilot filter installation. *(Courtesy Neptune Microfloc, Inc.)*

Typically, the pilot filters are contained in a console unit or in the plant control panel which may also house turbidimeters for monitoring the plant-scale filter effluent turbidity (see Figure 4-3).

Currently, two manufacturers (Neptune Microfloc and Turbitrol Co.) offer commercial pilot filter systems, both based on the same principal but differing somewhat in mechanical aspects. The Microfloc unit contains two, 4 1/2 in. diameter cylindrical, pressure pilot filters. Figures 4-4, 5, and 6 illustrate this system. Table 4-1 is a list of installations of this system. The Turbitrol pilot filters are placed in a tapered filter column (see Figure 4-7) to reduce the problems of media bridging the pilot columns during backwash. Also, this system uses an open top gravity filter to reduce the number of valves required. As the exact

TABLE 4-1 COAGULATION CONTROL SYSTEMS USING PILOT FILTERS

Water Treatment Plant Name	Plant Capacity, mgd	Location	Date of Installation
Hayden Bridge	81	Eugene, Ore.	1959
Arvada WTP	9	Arvada, Colo.	1961
Richland WTP	15	Richland, Wash.	1962
Silverton WTP	4	Silverton, Ore.	1962
Poplar Bluff WTP	5	Poplar Bluff, Mo.	1962
Pacific P & L Co. WTP	6	Green River, Wy.	1963
Silver City WTP	2.5	Silver City, N.C.	1963
Ute Conservancy Dist. WTP	4	Palisades, Colo.	1963
Clackamas WTP	6	Clackamas, Ore.	1963
Spiro WTP	1	Spiro, Okla.	1964
Craig WTP	5	Craig, Colo.	1964
Crown Zellerback Wauna Phase I WTP	15.1	Wauna, Ore.	1964
Stanton WTP	16	Stanton, Del.	1964
La Mariposa WTP	27	Caracas, Venezuela	1964
Pacific P & L Co. WTP	7	Albany, Ore.	1965
Greenville WTP	16	Greenville, Tex.	1966
Aurora WTP	20	Aurora, Colo.	1967
Longview WTP	13	Longview, Tex.	1966
Prince Albert	13	Prince Albert, Sask., Canada	1967
Medford WTP	15	Medord, Ore.	1967
Grants Pass WTP	11.52	Grants Pass, Ore.	1967
Palestine WTP	10	Palestine, Tex.	1967
Wilkesboro WTP	6	Wilkesboro, N.C.	1967
The Dalles WTP	4.3	The Dalles, Ore.	1968
Columbia WTP	12.6	Columbia, Tenn.	1968
Bellingham WTP	24	Bellingham, Wash.	1968
Holston defense area	11.52	Kingsport, Tenn.	1968
American Can Co.	18	Halsey, Ore.	1968
Naches River WTP	17.97	Yakima, Wash.	1968
Georgia-Pacific	45	Crossett, Ark.	1968
Beamsville	1.25	Beamsville, Ontario, Canada	1968
Orange Alamance	1	Mebane, N.C.	1968
Lake Oswego	10.3	Lake Oswego, Ore.	1968
Johnson City WTP	8	Johnson City, Tenn.	1969
Emporia WTP	4	Emporia, Va.	1969
Abilene WTP	16	Abilene, Tex.	1969
Newport WTP	—	Newport, Ore.	1969
Fremont WTP	7.5	Fremont, Ohio	1969
Douglas Taylor WTP	20	Corvallis, Ore.	1969
Fleming Hill WTP	24	Vallejo, Calif.	1969
Chambersburg MWTP	6	Chambersburg, Pa.	1969
Kenton County WTP	.26	Latonia, Ky.	1969
Longview MWTP	19.4	Longview, Tex.	1969

TABLE 4-1 (Continued)

Water Treatment Plant Name	Plant Capacity, mgd	Location	Date of Installation
Springfield MWTP	10	Springfield, Mo.	1969
Harrisonburg MWTP	5	Harrisonburg, Va.	1969
Anacortes MWTP	30	Anacortes, Wash.	1969
Harlingen WTP	6	Harlingen, Tex.	1970
Lexington Water Co.	5	Lexington, Ky.	1970
Chattanooga WTP	50	Chattanooga, Tenn.	1970
Caribou WTP	2.6	Caribou, Me.	1970
Lebanon WTP	10	Lebanon, Pa.	1970
American Oil Co.	36	Texas City, Tex.	1970

volume of flow through the filter is not critical, an excess of flow to the gravity filter is maintained at all times with the excess overflowing to a trough and then to waste as shown in Figure 4-7.

The pilot filter system can be used to accurately indicate coagulation conditions with a minimum lag time so that the operator can make the necessary adjustments in the plant chemical feed. Alternately, the coagulant feed can be controlled automatically.

Design for Automated Coagulant Control

This system is used in conjunction with a pilot filter system and automatically varies the plant coagulant dosage to maintain the effluent turbidity from the pilot filter at the desired set point value regardless of variations in raw water quality and other related factors. Other chemicals can be varied automatically by this system in direct relation to the coagulant dosage. The coagulant control system normally consists of a pilot filter system, an automatic control unit (see Figure 4-8), and switches and controls for the chemical feed equipment. The output signal from the automatic control unit to the chemical feeders can be a time duration, current, pneumatic, or other standard instrumentation control signal. In general, the time duration signal is found to be reliable and economic and is used quite extensively.

Basically, this automated control is accomplished by comparing a 0–10 JTU signal from the pilot filter turbidimeter with the desired turbidity value. If the turbidity signal is greater than the desired turbidity, the alum dosage will be increased. If the input turbidity is less than the set point turbidity, the alum dosage is automatically decreased. The plant operator establishes the desired quality through a set point potentiometer, and the unit adjusts the coagulant dosage to maintain the set point value. This unit continuously optimizes the alum dosage for a

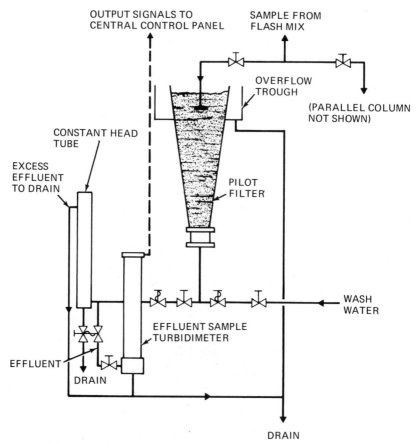

Fig. 4-7   Schematic diagram of a gravity pilot filter system. *(Courtesy Turbitrol Co.)*

given water treatment condition and consequently is constantly changing to establish the precise dosage. Figure 4-8 illustrates the control
panel of an automatic coagulant control system.

The system may incorporate provisions for automatic plant flow pacing. To allow for flow pacing, a potentiometer must be connected into
the plant raw water flow measurement equipment. This potentiometer
is integrated into the plant flow measurement system such that the
potentiometer wiper will be at 0 resistance with no flow and at maximum resistance at maximum plant flow. When the flow pacing feature
is incorporated, the dosage output will be automatically adjusted for
changes in plant flow. At the same time, changes in coagulant feed
dosage requirements will be maintained by the control unit.

The system may be provided with override means so that the plant
coagulant feed system may be operated in several modes. Four position

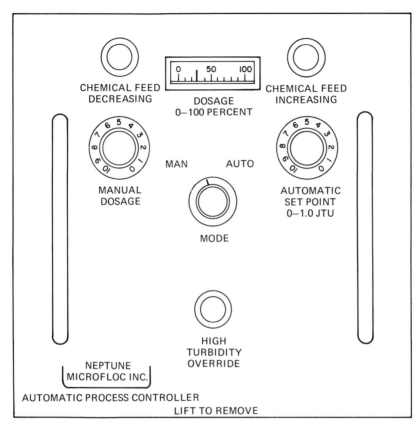

Fig. 4-8 Control panel of an automatic coagulant control system. *(Courtesy Neptune Microfloc, Inc.)*

selector switches are mounted on the face of the control panel for each chemical feeder with "Manual," "Off," "Plant Flow Pacing," and "Automatic Process Controller," positions. The operator may choose the operating mode for the coagulant feed system at any given time.

In the event of a control malfunction, the control unit will automatically provide a fixed chemical feed dosage to the plant raw water. This fixed dosage level is internally adjustable and is established based on the maximum average dosage required for the particular water. When this condition occurs, the high turbidity signal received causes the control unit to automatically switch to the fixed dosage value. At the same time the alarm light on the face of the control unit is energized. If, in the course of normal operation, the high turbidity value drops below the set point value of the fixed dosage control, the unit will again resume control of the plant chemical feed system on an automatic basis.

The chemical characteristics of the raw water will have an affect upon the adequacy of alum coagulation. Thus, all other variables which could affect the pilot filter turbidity must be controlled. The most important variable is pH. If the pH of the coagulated water is not maintained in the proper range for optimum filtration, the unit cannot function properly.

Waters with low natural alkalinity will generally require addition of artificial alkalinity to maintain the coagulated water pH at an optimum value for filtration. In most cases, the alkalinity requirement is a direct function of the coagulant dosage and the chemical feed pumps for both solutions can be proportioned over the entire range of expected coagulant dosages, supplementary artificial alkalinity feed is not required.

Filter Aid Control By Interface Monitoring

This technique has been developed in conjunction with dual-media filters. A sample of water is removed from the filter at a point near the interface between the coarse anthracite coal and the fine sand. The tubidity of this sample is continuously monitored and recorded, as is the filter effluent turbidity. Figure 4-9 shows the interface sampling system. The sample at the interface is obtained with a device which is a specially constructed well screen with slits small enough so that the coal and sand cannot pass through. The interface screen is placed 2 in. above the fine sand prior to anthracite placement. The turbidimeter is placed as close as possible to the filter. Samples flow by gravity to the turbidimeters. This prevents air bubbles which frequently occur when turbidimeter samples are pumped and also provides for the fastest possible response time. The interface turbidity sample is designed as a tool to aid the treatment plant operator in obtaining optimum performance from a dual media filter. The turbidity value obtained at the interface indicates where floc removal is occurring within the filter bed. For instance, a high interface turbidity and a low filter effluent turbidity indicates the proper coagulant dosage is being used but that significant amounts of floc are penetrating through the anthracite layer and must be removed by the fine sand. If the sand is forced to carry too much of the load, the headloss buildup becomes excessive, or breakthrough of turbidity could occur into the effluent. When floc removal does not occur in the anthracite layer when the proper coagulant dose is applied, the floc strength is inadequate and should be increased by the feed of polymers. This method works with any material which can be successfully used as a filter aid or conditioner. The filter aid must be started at a known underdose and incrementally increased until the desired floc removal is obtained in the coal layer. If an overdose is applied, it is usually impossible to use the interface concept until the filter has been washed.

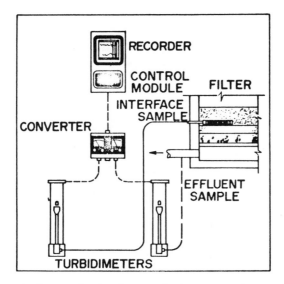

Fig. 4-9 Schematic diagram of an interface monitoring system. *(Courtesy Turbitrol Co.)*

It may also be desirable to adjust the filter aid dosage during the filter run, beginning with no aid or a small amount, and increasing the dosage as the run progresses.

TURBIDITY MONITORING

With the currently available, low cost turbidimeters, which are accurate and require little maintenance, every municipal water plant in this country, regardless of size, should provide a continuous record of the quality of its final product. Turbidity is an important parameter in that it reflects the efficiency of the coagulation process and the overall treatment provided, is related to probability of escape of pathogenic organisms from the treatment plant, and is a sensitive indicator of the aesthetic acceptability of the product to the consumer. An appalling number of plants provide no continuous monitoring of plant effluent turbidity. Visual estimates based on seeing the clearwell floor or plant log entries of "zero" turbidity at all times have been witnessed by the authors at many plants. There are very few production-oriented industries which do not maintain better quality control procedures than the water treatment industry, the performance of which is directly related to the public health of the nation.

For the measurement of very small amounts of turbidity, the principle of light scattering (nephelometry) is used. In this process, light is reflected at right angles to the light beam by the particles of turbidity in

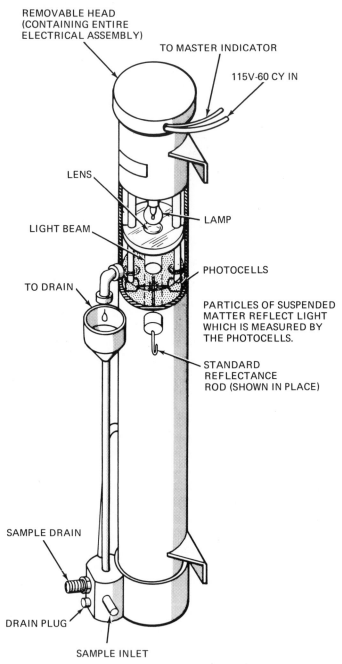

REMOVABLE HEAD
(CONTAINING ENTIRE
ELECTRICAL ASSEMBLY)

TO MASTER INDICATOR

115V-60 CY IN

LENS

LIGHT BEAM

LAMP

TO DRAIN

PHOTOCELLS

PARTICLES OF SUSPENDED
MATTER REFLECT LIGHT
WHICH IS MEASURED BY
THE PHOTOCELLS.

STANDARD
REFLECTANCE
ROD (SHOWN IN PLACE)

SAMPLE DRAIN

DRAIN PLUG

SAMPLE INLET

Fig. 4-10 Cross section of continuous flow, low range turbidimeter. *(Courtesy Hach Chemical Co.)*

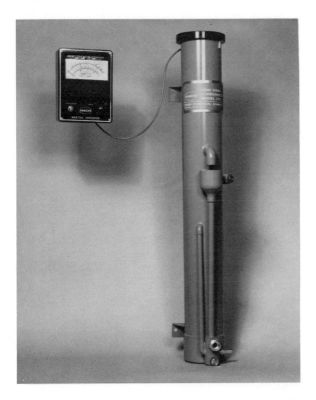

Fig. 4-11 Continuous flow, low range turbidimeter and indicating device. *(Courtesy Hach Chemical Co.)*

the liquid, as the beam of light passes through the liquid. The amount of reflected (scattered) light depends directly upon the amount of turbidity present in the liquid. This is similar to the common phenomenon of sunlight streaming through a window being reflected by otherwise invisible dust particles in the air.

If the liquid is entirely free of particles of turbidity, no scattered light will reach the photocells, and the indicating meter will read zero; thus, increasing turbidity gives an increase in meter reading.

The advantageous features of this system of turbidity measurement are:

1. A very strong light source can be used which permits a high degree of sensitivity; thus very small amounts of turbidity can be measured accurately.
2. The meter or output from the instrument is zero when the turbidity is zero.

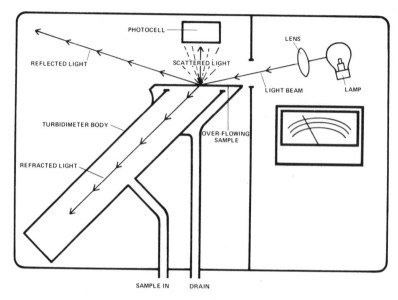

Fig. 4-12   Schematic diagram of Surface Scatter turbidimeter. *(Courtesy Hach Chemical Co.)*

Fig. 4-13   Surface Scatter Turbidimeter. *(Courtesy Hach Chemical Co.)*

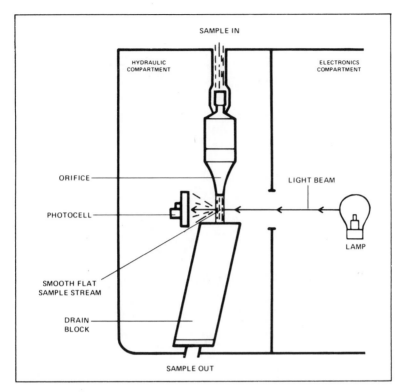

SAMPLE IN

HYDRAULIC
COMPARTMENT

ELECTRONICS
COMPARTMENT

ORIFICE

LIGHT BEAM

PHOTOCELL

LAMP

SMOOTH FLAT
SAMPLE STREAM

DRAIN
BLOCK

SAMPLE OUT

Fig. 4-14   Schematic diagram of a falling stream turbidimeter for measuring high turbidities. *(Courtesy Hach Chemical Co.)*

Units of this type (such as manufactured by Hach Chemical Company) are available with ranges of 0–0.2, 0–1, 0–3, and 0–30 turbidity units. A water sample of approximately 0.5 gpm at a head of 6 in. of water is required. Figures 4-10 and 4-11 illustrate this type of equipment which is best suited for measuring filter effluent turbidity or for interface turbidity monitoring.

The principle of nephelometry is the best system for turbidity measurement although it was once limited to low and medium range measurement because in the high ranges, with the conventional instrument, the light beam itself becomes absorbed by the high turbidity so that the instrument is unresponsive. In order to employ the light scattering principle without limit with respect to high range turbidities, a design employing "Surface Scatter" has been developed as illustrated in Figures 4-12 and 4-13. A very narrow beam is directed onto the surface of the liquid at an angle of 15 deg. Part of the beam is reflected by the water surface and escapes to a light trap. The remaining portion enters the water at approximately a 45 deg angle. If particles of

Fig. 4-15   Falling stream turbidimeter. *(Courtesy Hach Chemical Co.)*

turbidity are present, light scattering will take place, and some of the scattered light will reach the photocell. With this "Surface Scatter" design, there is no upper limit in the amount of turbidity it is possible to measure.

This unit is calibrated by using a Jackson candle turbidimeter to measure the turbidity of the same water that is flowing through the instrument. The instrument is adjusted so that its output reading corresponds to the Jackson candle reading. Ranges as high as 0–5000 are available and this instrument is used for measuring the turbidity of raw water or settling basin effluent. A 0.25–0.50 gpm sample stream is required.

High turbidities (250 JTU and above) may be measured by using the light transmitted through a falling stream of the sample, as shown in Figures 4-14 and 4-15. The turbidimeter is unique in that no part of the system optics is in contact with the sample, reducing cleaning and maintenance. A 1 gpm sample stream is required. The unit is used for measuring the sludge blanket location in a clarifier or other uses involving very high turbidities.

REFERENCES

1. TeKippe, R. J., and Ham, R. K., "Coagulation Testing, A Comparison of Techniques," *Journal American Water Works Association*, **62**, Part I, P. 594; Part II, p. 620 (1970).
2. Conley, W. R., and Evers, R. H., "Coagulation Control," *Journal American Water Works Association*, **60**, p. 165 (1968).
3. Black, A. P., and Smith, A. L., "Determination of the Mobility of Colloidal Particles by Microelectrophoresis," *Journal American Water Works Association*, **54**, p. 926 (1962).
4. Williams, R. L., "Microelectrophoretic Studies of Coagulation With Aluminum Sulfate," *Journal American Water Works Association*, **57**, p. 801 (1965).
5. Riddick, T. M., "Zeta Potential and Its Application to Difficult Waters," *Journal American Water Works Association*, **53**, p. 1007 (Aug. 1961).
6. Shull, K. E., "Filtrability Techniques for Improving Water Clarification," *Journal American Water Works Association*, **59**, p.1164 (1967).
7. Kawamura, S., and Tanaka, Y., "Applying Colloid Titration Techniques to Coagulant Dosage Control," *Water & Sewage Works*, **113**, p. 348 (1966).
8. Kawamura, S., Sanna, G. P., Jr., & Shumate, K. S., "Application of Colloid Titration Technique to Flocculation Control," *Journal American Water Works Association*, **59**, p. 1003 (1967).
9. Harris, R. H., "Control of Filter Aids in High Rate Filters Using Turbidity Monitoring," presented at the New York American Water Works Association meeting (October 8, 1970).
10. Gerdes, W. F., "A New Instrument — The Streaming Current Detector," unpublished.
11. "Information Resource: Water Pollution Control in the Water Utility Industry," American Water Works Association Research Foundation, EPA Report 12120 EUR 11/71 (November, 1971).

5 Disposal of sludges

NATURE OF THE PROBLEM

The predominant method of disposal of backwash waste-waters and sludges from water treatment plants has been and continues to be direct discharge to surface waters. A survey of plants in 1953[1] showed that 92.5 percent of the plants discharged sludges directly to the receiving stream while a 1968 study[2] showed only a small percentage of the plants using lagoons or other available methods of sludge disposal. Yet a 1968 survey of state pollution control agencies[3] reported that 43 of the 47 responding states considered discharge of these wastes to surface waters a violation of pollution laws or regulations. It is apparent that current pollution control programs and past water treatment industry practice are on a collision course on this topic.

The wastes generated by a water treatment plant are composed of the solids removed from the raw water as well as the chemical solids resulting from the chemical treatment of the raw water. The nature and quantity of the raw water solids obviously will vary drastically from one plant to another but also may vary substantially even at a given plant due to variations in sediment washed into the water supply by rainfall, seasonal algal growths, spring turnover of a lake, etc. The nature and quantity of chemical solids are primarily a function of the purpose of the coagulation step: i.e., turbidity removal, iron or manganese removal, or hardness removal. The gelatinous hydroxide flocs

136

resulting from the addition of aluminum salts are difficult to dewater while floc resulting from lime additions for softening are more readily dewatered.

The wastewaters originate from two sources within the plant: the clarifier underflow and the filter backwash wastewaters. The great majority of the solids will be found in the clarifier underflow (assuming the clarifier is operating properly) with only a small fraction of the solids being found in the filter backwash wastewaters. The volume of filter backwash wastewaters is generally 5–10 times as great as the underflow volume from the clarifier and represents a very dilute wastewater. Many plants now recycle filter backwash waters to the head of the plant so that only one concentrated source of waste solids must be handled — the clarifier underflow.

Although the wastewaters from a water treatment plant contain primary solids removed from the water source (as well as the chemical solids, of course), they are potential pollutants because they may produce objectionable sludge deposits in the receiving water which are aesthetically objectionable or which may form deposits harmful to bottom organisms in the receiving stream. Generally, the settleability of the material removed from the water is increased in the water treatment plant so that when it is returned to the same watercourse, it is much more likely to form a sludge deposit.

The fundamental problem in disposal of water treatment plant sludges is to concentrate them and render them innocuous so that they are suitable for disposal at sites other than in a water supply. Filter backwash wastes are far too dilute to subject to a dewatering step without first separating the solids from the washwater by gravity. This separation can either be achieved by recycling the backwash wastewaters to the plant clarifier or by routing the backwash wastes through a separate set of clarifiers[5] and combining the sludges from both sets of clarifiers for additional treatment. This chapter will discuss the alternates available for handling the sludges resulting from sedimentation of coagulated waters and from sedimentation of filter backwash wastes. Although the techniques described in this chapter are not new, their application to water treatment sludges is largely new. The past application to water sludges have been scattered and few. The lack of characterization of the sludges involved in these past applications makes derivation of detailed design guidelines difficult, as will become apparent as this chapter is presented.

Sludge Characteristics

Obviously, sludge characteristics vary widely as do the characteristics of the waters being treated through the country. Published data[3,6] indicate sludges resulting from alum coagulation have

BOD values of 40–150 mg/l, COD values of 340–5000 mg/l, total suspended solids of 1100–14,000 mg/l (values of 5000–10,000 predominate), volatile suspended solids of 600–4000 mg/l (generally 30–50 percent volatile), and a pH near neutral. The dry weight of the sludges is typically 75–95 lb/ft³. Alum sludge at 20 percent solids has been reported to have a unit weight of 73 lb/ft³.

The major constituent resulting from lime coagulation of wastewaters is calcium carbonate which generally makes up 85 to 95 percent of the dry weight of solids in the sludge from such a plant. The quantities of calcium carbonate produced during lime treatment are quite substantial, as 1 lb of pure quicklime (calcium oxide) when added to water will react with the calcium bicarbonate in the water to precipitate 3.57 lb of calcium carbonate. As commercial quicklime usually contains only 85 to 95 percent calcium oxide, and as the lime also reacts with some of the magnesium, iron, and aluminum that may be present, the amount of sludge produced during lime treatment is generally assumed to be about 2.5 lb, dry basis, for each lb of quicklime added.[3]

ALTERNATES FOR DISPOSAL

Discharge to Sanitary Sewers

Comparison of the 1953 and 1968 surveys discussed earlier shows a marked increase in the percentage of plants discharging their sludges to sanitary sewers, 0.3 percent in 1953 and 8.3 percent in 1968. This technique merely transfers the solids handling problem from the water treatment plant to the waste treatment plant. However, inclusion of the necessary capabilities in the solids handling facilities of the waste treatment facility may result in an overall cost savings by consolidating the equipment and reducing the number of personnel required for the total solids handling problem. There are a number of considerations which require evaluation if this approach receives serious consideration for a given application.

A major consideration is the ability of the waste treatment plant to accept the increased hydraulic and solids load imposed by the addition of the water treatment plant wastes. The direct injection of the filter backwash flows into the sewer system, for example, could cause a hydraulic surge enough to cause the waste treatment plant clarifier performance to deteriorate. Surge storage at the water plant may be needed so that the volume of water plant wastes, if large in proportion to the sewage flows, may be released at a controlled rate. Release during low sewage flow periods (midnight–6 a.m.) may be desirable.

The bulk of the solids from the water plant sludges will be removed in the primary clarifier. Obviously, the solids handling system at the waste treatment plant must be capable of handling the additional solids load. It will probably be a rare but fortuitous circumstance when an existing waste treatment plant solids handling system can handle the unplanned addition of water plant sludges if the water plant and waste plants are of comparable size. For example, an injection of large amounts of gelatinous hydroxide floc into an anaerobic digestion–sand drying bed system may cause difficulties in obtaining proper solids dewatering. A careful evaluation of the dewatering characteristics of the combined water and sewage sludges should be made prior to adopting this approach.

Another aspect to consider is that the sewer receiving the water plant sludges is of adequate capacity and will provide velocities adequate to prevent deposition of the sludge in the sewer. Studies at Detroit[3] report a velocity of 2.5 fps is adequate to prevent settling of the sludges in the sewer.

Successful injection of lime sludges into the Daytona Beach, Florida sewer system have been reported.[3] About 15,700 lb (dry solids) of sludge was added to a sewage treatment plant sized for 70,000 people with the resulting sludges being dewatered by a vacuum filter. At the time of this writing, Dallas, Texas was beginning operation of a system for injection of lime sludges into a system to treat combined sewer overflows. The large quantities of inert, dense materials found in softening sludges may lead to filling of anaerobic digesters at the waste treatment plant.

Lagoons

The most common method now used for handling of water plant sludges is lagooning. The operating costs of this technique are low but the land requirements are high. Because of the space requirement, lagooning may be more attractive for small, isolated plants.

Lagoons are generally built by enclosure of a land area by dikes or excavation with no attempt to maximize drainage with underdrains or by a sand layer. Sludge is added until the lagoon is filled with solids at which time it is removed from service until the solids have dried to the point that they can be removed for landfill disposal, if in fact the sludge can dewater to this point. Lagooning is not so much a disposal technique as it is one for thickening, dewatering, and temporary storage.

Alum sludges have proven difficult to dewater in lagoons to the point that they can be handled for landfill. Neubauer[7] reported a detailed study of a lagoon receiving alum sludges from a 32 mgd plant in Rochester, New York. The 400 ft long by 320 ft wide lagoon had been in operation for 3 years. Core samples indicated that the solids concentrated increased from about 1.7 percent at the sludge interface to a maximum of 14 percent at the lagoon bottom. The average solids concentration of the sludge was 4.3 percent with a majority of the sludge having less than 10 percent solids concentration. It was concluded that the lagoon did not produce a sludge suitable for landfill disposal without further dewatering. Other plants[3] report removing the thickened alum sludge by dragline or clamshell for dumping the sludge in thin layers on the lagoon banks to air dry; dumping the thickened sludge on land disposal areas or on roadsides; or transporting the thickened sludge to a specially prepared drying bed. The Somerville, New Jersey[3] plant which has a capacity of 170 mgd, offers an example of circumstances in which lagooning of alum sludges has provided a satisfactory solution to the sludge handling problem. A low turbidity raw water (1–30 Turbidity Units) receives an alum dose of 40 mg/l prior to sedimentation and rapid sand filtration. The sludge is removed twice per year from the sedimentation basin and pumped to a lagoon 400 × 1200 ft in area (total land used = 12 acres) formed by a previously existing roadway fill on three sides and a dike on the fourth. Supernatant is decanted from the lagoons by overflow pipes. The total available sludge storage depth of 6 ft had filled to a depth of 3 1/2 feet after 10 years of operation. The sludge was dry enough to permit walking and was estimated to have a solids content of 30 percent. It was estimated that the lagoon would have a total useful life of 20 years with slight additions to the dike.

In climates where freezing temperatures occur frequently, the freezing and thawing of lagooned alum sludges may result in a marked improvement in the dewatering of the sludges. Freezing has been reported[2] to result in an increase in total solids up to 17.5 percent. Farrell et al.[8] report that a freezing-thawing cycle increased the solids in an alum sludge from 3.2 percent to 17.6 percent. They found that snow cover was a serious obstacle which could best be overcome by freezing the lagooned sludge in thin layers, by not putting out sludge when snow is expected, by melting the snow with a water flush, or by layering out more sludge. Freezing in the Cincinnati, Ohio climate was successful only if thin layers of sludge (on the order of an inch) were applied at a time which leads to large space requirements. Tests conducted in New York State[4] indicated that 0.3 percent solids sludge placed in a lagoon in January with a depth of 30 in. which was subjected to natural freezing dewatered to 35 percent solids by the next August by decanting the liquid. Allowing the sludge remaining after decanting to stand for one week in 80°F weather then reduced the solids content to about 50 percent, suitable for handling and disposal as land fill. The only full-scale natural freezing operation involving alum sludge is being practiced by the city of Copenhagen, Denmark[3]. The lagoons, in use since 1963, place a depth of sludge of 1 meter in a lagoon with a drained gravel bottom. As a result of the freezing and thawing cycles and drainage through the gravel, the sludge concentration is very significant. It was reported that 2 years of operation left only a 4 cm deep layer of sludge.

Lime sludges are more easily dewatered in lagoons than are alum sludges. Several reports indicate that a solids content of 50 percent may be achieved with lime sludges, which is adequate for handling and disposal in a sanitary land fill operation. Based upon a 50 percent solids content, the lagoon capacity requirements are about 0.5–0.7 acre-ft/year/mgd/100 mg/l hardness removed. Lagoon depths are usually 3–10 feet. The 10 State Standards recommend a minimum lagoon depth of 4–5 ft, a capacity of 3–5 years solids storage with multiple cells and adjustable decanting devices. Lime sludges were used to fill in low land areas around the Miami, Florida water treatment plant and were later covered with 1–2 in. of soil and sown with grass.[3] However, lagooned lime sludges are generally considered to be a poor fill material[3] and final disposal of the dried material may still present a problem.

Serial lagoons are used at the Willingboro, New Jersey plant for a sludge resulting from the addition of 32 mg/l lime and 5 mg/l alum to an average flow of 2.1 mgd.[3] The sludge is drained into 2 parallel thickening lagoons which are 75 ft × 75 ft × 9 ft deep. Once per day the thickening lagoons are decanted. About 3 times per year, the sludge is pumped from the thickening lagoons to a drying lagoon which is 75 ft × 100 ft × 3 ft deep which is built with a gravel base covered by sand. About every 2 years the sludge is removed by front end loader from the drying lagoon at a solids content of about 30 percent. Costs were estimated at about $40/ton of dry solids.

Sand Drying Beds

Sand beds, similar to those used for drying of sewage sludges, have also been used for drying water plant sludges. These beds usually consist of 6–9 in. of sand over a 12 in. deep gravel underdrain system made up of 1/8–1/4 in. graded gravel overlying drain tiles. The rate at which sludges placed on such beds will dewater depends upon the air temperature and humidity, wind currents, and viscosity of the sludge. Sand sizes of about 0.5 mm are typically used with a uniformity coefficient of less than 5. Excessively coarse sands result in too great a loss of solids in the drying bed filtrate. Neubauer[7] reports that with a 5 mile per hr wind, over ranges of temperatures of 69–81°F and humidities of 72–93 percent, solids concentrations of 20 percent were achieved from an alum sludge in 70–100 hr with 97 percent capture of solids with a solids loading of 0.8 lb/ft². Use of sand sizes of 0.38, 0.50, and 0.66 mm effective size made little difference in total drying time. Alum sludges are dewatered on drying beds at the Sunol Valley Filtration Plant in California.[3] Alum coagulation is practiced only during the winter season and sludge is continuously added to the drying beds during the coagulation season. Supernatant is decanted from the drying beds. The depth of sludge added to the beds has had a major effect on the results. With a sludge depth of 4 ft, a 6 month drying period resulted in only the upper 2 ft of sludge drying and cracking. A 2 ft sludge layer dried in 4 months to the point that it could be removed by a front end loader for hauling by truck to a disposal site on water department property. The cost was estimated at about $60/ton of dry solids. Lime sludges have been reported [3] to dewater to 50 percent solids in 4 months time on drying beds in Lompoc, California when placed on the bed in a 5 ft deep layer. The drying beds at Lompoc were constructed by placing a 15 ft diameter, 15 ft deep sand core in the center of the 140 ft square drying beds to aid in draining moisture from the sludge. The drying beds are filled with sludge and then decanted with the decant returned to the plant influent. The dewatered lime sludge is removed with front end loaders and hauled by truck to a landfill disposal site. The total costs for the Lompoc system for lime sludges are estimated at about $5/ton.[3]

The use of sand drying beds has the following inherent disadvantages: requires extensive space and long dewatering times, poor to inadequate performance during cold or wet periods, and high labor costs for removal of sludge. The mechanical dewatering devices described in this chapter offer a more positive, more controllable, and more compact means of dewatering sludges.

Vacuum Filtration

A vacuum filter consists of a suspended drum which rotates in a container of sludge with about one fourth of the outer surface of the drum being submerged. Sections of the drum are alternately subjected to suction and pressure during each revolution. The surface of the drum is covered with a filter medium through which the water is drawn as the drum passes through the sludge. The solids are deposited as a cake on the filter medium. As this cake rises out of the liquid sludge, air is drawn through the cake for further dewatering. The cake is then scraped off the filter medium before it is again submerged to pick up a new layer of sludge.

The filter medium used has an important bearing on vacuum filter performance. The two general types available are coil springs and cloth. The Komline-Sanderson Coilfilter uses two layers of coil springs. The filtrate passes through and between the coils. As the drum revolves, one layer of the coils separates from the other and conveys the dry sludge cake to a point of discharge from the machine. The coils are then washed before making another revolution around the drum. There are a number of vacuum filters which use cloth or fabric media. Wool, cotton, orlon, nylon, Dacron, polyester fiber, and other materials are used. Vacuum filters with cloth or fabric medium are supplied by Komline-Sanderson, Dorr-Oliver, Walter Process, Eimco, Rex Chainbelt, and other companies.

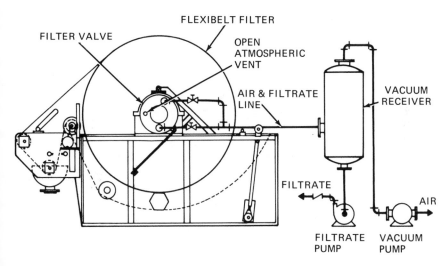

Fig. 5-1 Schematic drawing of typical vacuum filter installation. (*Courtesy Komline-Sanderson Corp.*)

A vacuum filter installation includes a sludge feed pump, chemical feeders, sludge conditioning tanks, vacuum filter, vacuum pump and receiving tank, filtrate pump, and filter cake conveyor and hopper. Figure 5-1 is a schematic diagram of a typical vacuum filter installation. It is sometimes necessary to precondition sludges or slurries to be dewatered with coagulating chemicals such as, ferric chloride, lime or polyelectrolytes, for example, or filter aids. This may be accomplished in a rotating conditioning tank (see Figure 5-2). Two chemical application points are provided and the rotating drum (variable speed) is equipped with an annular baffle, producing isolated mixing chambers and nonclog mixing blades.

Vacuum filters are available in a wide range of sizes with diameters of 3 to 12 ft and lengths of 1 to 20 ft. Table 5-1 provides a summary of the filtering area associated with a variety of vacuum filter sizes.

Although vacuum filters are widely used in dewatering sewage sludges, they have had limited application in the water treatment field. Alum sludges have proven difficult to dewater by vacuum filtration (or by any other means for that matter). The gelatinous nature of alum sludges almost precludes the use of vacuum filtration without precoating the filter with diatomaceous earth. The cost of precoating is high, however. Also, the remaining sludge–precoat mixture remains gelatinous in nature and the combination may not be well-suited for ultimate disposal because the residual moisture content may be retained[4] although tests at Albany, New York discussed below indicate this need not always be the case. Solids concentrations as high as 20 percent have been reported to be achieved by precoated vacuum filters.

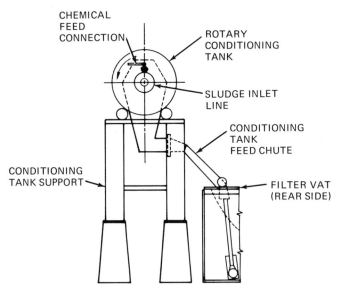

Fig. 5-2 Schematic drawing of a sludge conditioning system. (*Courtesy Komline-Sanderson Corp.*)

The rotary-vacuum precoat filter (RVPF) is a modification of a continuous rotary-vacuum filter, which permits filtering of materials higher in suspended solids than can be handled by other means. RVPF provides long (6–40 hr) filtration cycles while reducing the problem of filter media "blinding." A precoat of filter aid 2–4 in. thick is applied to the filter by introducing a dilute filter aid into the filter bowl and applying vacuum to the rotating filter drum.

After the precoat is in place the sludge is introduced into the bowl and the filter cycle is started. The precoated drum is submerged in the

TABLE 5-1 FILTERING AREA OF VACUUM FILTERS

Filtering Area of Drum Filters in Square Feet

Diameter	Face (in ft)																			
	1	2	3	4	5	6	7	8	9	10	11	12	13	14	15	16	17	18	19	20
3'	9.4	18.8	28.2	37.7	47.1	56.5														
4½'		—	42.4	56.5	70.6	84.8	98.9	103	127											
6'				—	94.2	113	132	151	170	188	207	226								
8'						—	176	201	226	251	277	302	327	352	377	402				
10'								—	283	314	345	377	409	440	471	502	534	565	596	628
12'										377	415	452	490	528	565	603	641	679	716	754

sludge to about 30–50 percent of its depth and is continuously rotated with constant vacuum applied to the cake.

Recent tests were conducted at Albany, New York on a sludge resulting from alum coagulation of a low turbidity water.[28] The sludges at Albany are difficult to dewater and are high (69 percent) in alum content. Results of tests conducted with a 36 in. diameter by 6 in. face width vacuum filter are summarized in Table 5-2. These results show that a 20 percent cake solids concentration was readily achieved at flow rates of 5 gal/ft²/hr. Filtrate clarity throughout the study was also excellent, averaging less than 5 JTU with greater than 99 percent suspended solids removal.

TABLE 5-2 CAKE SOLIDS CONTENT AND FILTRATE TURBIDITIES VACUUM FILTRATION WITH VARIOUS PRECOATS[28]

Precoat	Drum Sub-mergence per cent	Drum Speed min/rev.	Sludge Solids per cent	Cake Solids per cent	Avg. Cake Solids per cent	Filtrate Turbidity JTU	Avg. Filtrate Tur-bidity JTU
Hyflo-Super-Cel[a]	58	1	0.68	26.2		3.8	
	58	1	0.48	30.1		3.2	
	58	2	0.69	24.5		4.4	
	58	2	0.48	31.0		4.8	
					29.02		4.07
	33	1	0.70	28.8		3.8	
	33	1	0.38	29.8		2.5	
	33	2	0.70	30.7		5.3	
	33	2	0.38	31.9		4.8	
Celite 501[a]	58	1	0.62	26.0		4.6	
	58	1	0.56	27.6		3.9	
	58	2	0.62	25.5		6.4	
	58	2	0.56	29.5		4.9	
					31.60		4.41
	33	1	0.78	26.1		3.3	
	33	1	0.50	30.0		3.6	
	33	2	0.78	27.2		4.5	
	33	2	0.50	31.9		4.1	
Celite 503[a]	58	1	0.64	25.8		4.4	
	58	1	0.43	28.8		4.2	
	58	2	0.64	26.5		3.2	
	58	2	0.43	30.4		5.7	
					29.44		3.72
	33	1	0.71	28.3		2.7	
	33	1	0.34	31.6		2.3	
	33	2	0.71	31.2		3.2	
	33	2	0.34	32.9		4.1	

[a]Johns-Manville Products Corp.

The filter cake was reported as suitable for easy handling as landfill. Polymer treatment of the sludge was reported as ineffective for improving filterability. It was estimated that application of the RVPF at Albany, New York would add $6.15/mg to the current treatment costs of $20.00/mg. These costs were equivalent to $11.15/1000 gal of alum sludge (0.7 percent solids).

Some investigators[7] have reported that polyelectrolytes were of little benefit in filtering alum sludges although others have reported an increase in loadings on the filter by use of the proper polymers. Polymer dosages as low as 0.10 percent by weight have been reported[25] to lower the specific resistance of alum sludge tenfold from 20.7 $sec^2/g \times 10^8$ to 2.25 $sec^2/g \times 10^8$. The dosage varied with the polymer used with the anionic and nonionic polymers requiring the lower dosages (0.14 and 0.26 percent by weight, respectively) to produce minimum specific resistance with cationic polymers requiring higher dosages of 1.5–4.8 percent by weight. The addition of lime to raise the pH of alum sludges to 10.5 has been reported to provide improved dewatering.[2]

Vacuum filtration of lime sludges has been more successful. Installations at Boca Raton, Florida and Minot, North Dakota both are reported to dewater thickened lime sludges of 30 percent solids to a cakes of 45–65 percent solids, adequate for use as a landfill.[3] Cloth belt lifes of 6–9 months were reported. At the Boca Raton plant, 120–130 mg/l lime is added to soften a water with 210–250 mg/l hardness to provide a 58 mg/l total hardness effluent. The raw water contains only 6 mg/l magnesium. Average plant flow was 7.6 mgd during the study period reported.[3] Sludge is removed continuously from upflow clarifiers at 1–4 percent solids content and pumped to a 25 ft diameter thickener. The sludge is low in magnesium hydroxide, making it more readily dewaterable. Thickened sludge (28–32 percent solids) is pumped to a vacuum filter which had been in use since 1964–65. The cake discharged consists of 65–70 percent solids and is directly used as a road stabilizer. Solids loadings of 60 lb of dry solids per hr per ft² are reported as the average operating condition. A small lagoon serves as an emergency sludge storage basin to be used when the vacuum filter is removed from service for repair. The solids production is about 1 ton/mg and the estimated cost, 70 percent of which is due to labor and supervision, was estimated at $16.00/ton (1968 prices). The Minot plant processes 10 mgd during summer months with the raw water having a high hardness (about 115 mg/l). About 3 tons of dry solids/mg are produced. The sludge passes through a thickener prior to vacuum filtration which produces a cake with 44 percent solids at loadings of 30 lb dry solids per hr per ft². The sludge cake is placed in a landfill on the plant site. Experiences with hauling to a town dump indicated that rubber seals in the dump truck box were necessary to minimize sludge

Fig. 5-3 Vacuum filter operating on lime sludge. *(Courtesy Komline-Sanderson Corp.)*

leakage during transport. Total costs were estimated at about $8/ton of dry solids, 50 percent of which were operation and maintenance costs. Figure 5-3 illustrates a vacuum filter in operation on lime sludges.

The AWWA has recently reported on model studies of vacuum filtration of both alum and lime sludges.[3] These studies were made on the basis of a 10 mgd plant providing coagulation, sedimentation, and rapid sand filtration with the sludges receiving thickening and vacuum filtration prior to land disposal. Table 5-3 summarizes the results of these studies. The difficulties encountered in dewatering alum sludges are readily apparent from this comparison.

Centrifugation

Centrifuges are widely applied to the dewatering and classification of water softening sludges with more than 150 installations in industrial and municipal applications reported.[9] Many applications to predominantly alum sludges have not been successful with maximum solids concentrations of only 12–15 percent being obtained,

TABLE 5-3   MODEL STUDIES OF VACUUM FILTRATION FOR A 10 mgd PLANT[3]

	Alum Sludges	Lime Sludges
Sludge quantities, gpd		
From clarifier	50,000	240,000
Sludge concentration, % solids		
From clarifier	1	1
From thickener	2	30
Thickener loading, lbs dry solids/ft²/day	4.0	12.5
Thickener diameter, ft	35	45
Vacuum filter loading, lb dry solids/ft²/hr	0.4[a]	40
Vacuum filter area required, ft²	1,600	94
Vacuum filters selected[b]		
Number	3	1
Size	14 × 18 ft	6 × 5 ft
Sludge concentration from vacuum filter, % solids	25	65
Estimated capital costs[c]	$609,400	$67,000
Estimated annual costs	150,600	45,000
Total costs		
$/mg	41.30	12.35
$/ton dry solids	198.00	12.35

[a]With precoat of diatomaceous earth.
[b]Assumes 6 hr operation of filters/day.
[c]1968 prices.

although one major installation at a 40 mgd plant using alum coagulation was underway at the time of this writing (Sobrante, California plant).

The centrifuge is essentially a sedimentation device in which the solids-liquid separation is enhanced by rotating the liquid at high speeds so as to subject the sludge to increased gravitational forces. The most suitable centrifuge design for lime sludges is the continuously discharging solid bowl machine. The principles on which this machine are based are illustrated in Figure 5-4. The sludge is introduced into the revolving bowl (Figure 5-5) through a stationary feed tube at the center of rotation. The solids are thrown against the wall of the bowl. The lighter liquid forms a concentric layer inside the solids layer in the bowl. Inside the bowl is a helical screw conveyor or "scroll" (Figure 5-6) which rotates in the same direction as the bowl but at a slightly lower speed. This conveyer moves the solids deposited against the bowl to one end of the bowl where they are "plowed" up the "beach" and out of the liquid layer where they are discharged from the bowl

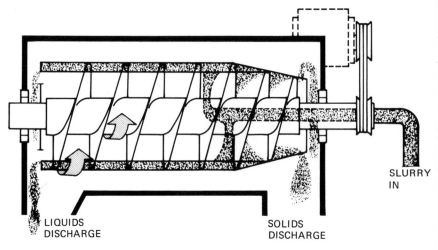

Fig. 5-4 Operational principal of a centrifuge.

through suitably located ports. The typical construction of a horizontal, solid bowl machine is shown in Figure 5-7. Ports in the bowl head act as overflow weirs for discharge of clarified effluent. The location of these ports with respect to the axis is adjustable and determines the level of slurry retained in the bowl.Usually this slurry level is set so that the liquid in the bowl submerges all but a portion of the conical drainage deck. A solid bowl centrifuge must carry out the dual func-

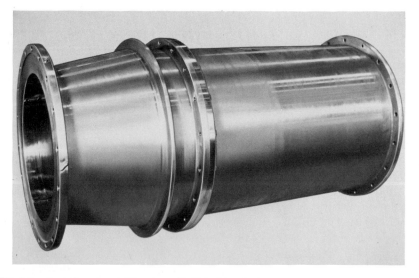

Fig. 5-5 Centrifuge bowl. *(Courtesy Bird Machine Co.)*

Fig. 5-6   Typical conveyor. *(Courtesy of Bird Machine Co.)*

tions of clarifying the incoming sludge and conveying the solids out of the bowl. Increasing the centrifugal force and lowering the liquid depth in the bowl, for example, will theoretically improve clarification but in many instances may act to the detriment of the machine by hindering the conveying of solids.

Most solid bowl machines employ the countercurrent flow of liquid and solids described above and illustrated in Figure 5-4 and are appropriately referred to as "countercurrent" centrifuges. Recently, a "cocurrent" centrifuge design has also been introduced which theoretically offers the following advantages: (1) the settled solids are not disturbed by the incoming feed, an inherent problem with the countercurrent design, (2) turbulence is reduced by having the solids and liquid flow in the same direction and (3) the solids have the entire length of the bowl to settle and compact, providing for better conveying up the beach. General construction is similar to the countercurrent design except there are no effluent ports in the bowl head. Instead, the clarified effluent is withdrawn by a skimming device near the junction of the bowl and the beach. The feed is introduced at the end opposite the beach in this type of machine. Parallel tests indicate that the concurrent flow design does result in improved performance. When separating a fine calcium carbonate floc with 30 percent solids concentration, the concurrent centrifuge was found to have about 60 percent greater capacity for an equal degree of clarification.[10] In general, the advantages of the concurrent design are pronounced when dealing with a high solids concentration or where the solids form a voluminous, low density cake.

The rating of centrifuge capacity is a complex problem due to the large number of operational variables. Generally, the capacity is expressed in terms of lb or gallons of solids removed per hr. Large machines may handle up to 100 tons of solids per hr. Centrifuge bowl

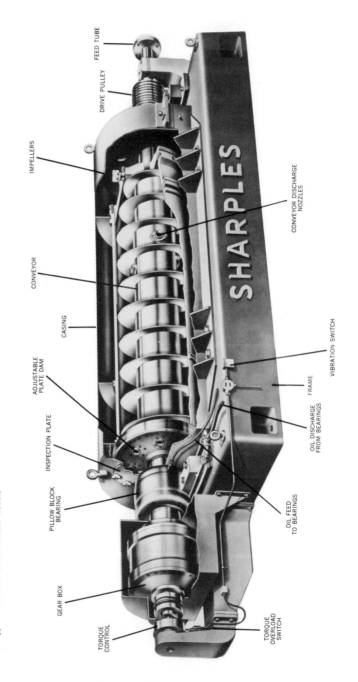

Cutaway view of Model P-5400 illustrating typical features of horizontal models

FEED TUBE

DRIVE PULLEY

IMPELLERS

CONVEYOR DISCHARGE NOZZLES

CONVEYOR

CASING

VIBRATION SWITCH

ADJUSTABLE PLATE DAM

FRAME

INSPECTION PLATE

OIL DISCHARGE FROM BEARINGS

PILLOW BLOCK BEARING

OIL FEED TO BEARINGS

GEAR BOX

TORQUE CONTROL

TORQUE OVERLOAD SWITCH

Fig. 5-7   Cross section of a horizontal, solid bowl centrifuge. (*Courtesy Sharples Co.*)

TABLE 5-4 APPROXIMATE CENTRIFUGE CAPACITIES FOR WATER SOFTENING SLUDGES[11]

Machine Size	Capacity, tons/hr	Space Required
18 in.	0.5–1.5	5 × 10 ft
24 in.	1.0–3.0	6 × 12 ft
36 in.	3.0–5.0	10 × 16 ft

sizes range from 18 in. in diameter to 54 in. in diameter. Horsepower requirements in this size range vary from 15 to 200 hp. Several controllable machines variables affect centrifuge performance. Higher through-put rates decrease solids recovery but increases cake solids. A greater pool depth improves recovery but leads to wetter cake. Conversely, a shallower depth decreases recovery but produces a dryer cake. Increased bowl speed will provide higher recovery and higher cake solids for lime sludges. General guides for water softening sludges are shown in Table 5-4. These capacities will vary with sludge characteristics and are very general estimates. Without flocculants, 70–90 percent solids recoveries with 50–65 percent solids concentrations in the cake are typically obtained. The use of polymers may increase the percent recovery to 95 percent but will generally decrease the cake solids to 40–55 percent.

Water softening applications of centrifuges may have one of two goals:

(1) Dewatering with the objective of maximum capture of suspended solids.
(2) Dewatering and classification to produce as pure a calcium carbonate cake as possible.

When lime is added to waters with a high magnesium content, large quantities of magnesium hydroxide are precipitated which renders the sludge more difficult to thicken and dewater. If lime recalcining is practiced, the magnesium oxide produced from this floc in the recalcining furnace will appear as recycled inert material which will result in an ever increasing amount of solids in the system. Another approach involves carbonation of the lime sludges with recalciner flue gas rich in CO_2 to convert magnesium hydroxide to dissolved magnesium carbonate. An alternate approach is to use a centrifuge to classify the sludge into its calcium carbonate and noncalcium carbonate components. This classification can be accomplished due to the difference in the specific gravities of calcium carbonate and the impurities. By proper operation, the lighter impurities can be permitted to exit in the centrate. Control of the pool depth permits this classification to be adjusted as desired while the machine is in operation. Preferably, the cake produced would be 100 percent calcium carbonate with 100 percent of the impurities rejected in the centrate. Recovery levels with good classification will be 75–85 percent of the suspended solids and the cake will be drier because the gelatinous materials, much in the form of magnesium hydroxide, are eliminated from the cake. The ability of the centrifuge to provide magnesium classification is a major advantage of centrifuges over vacuum filters when recalcining is practiced.

A recent AWWA study reported on experiences with centrifuges at four large plants: Austin, Texas; Dayton, Ohio; Lansing, Michigan; and Miami, Florida. All 4 plants used 40 in. diameter × 60 in. long solid bowl centrigures. Table 5-5 summarizes the operating and cost data from these 4 plants. A model study similar to that discussed earlier for vacuum filtration was made on centrifucation of lime sludge following thickening.[3] The cost for a facility to handle the same lime sludge shown in Table 5-3 was estimated at $98,000 with annual costs of $41,350 for a cost of $11.40/ton of dry solids and $11.40/mg water treated. Costs for vacuum filtration were shown as $12.35/ton in Table 5-3 for the same assumed conditions.

TABLE 5-5 DATA FROM ILLUSTRATIVE CENTRIFUGE INSTALLATIONS ON LIME SLUDGES[3]

Capacity, mgd	120	120	40	120
Ave flow, mgd	30	61.2	23	81
Influent hardness, mg/l as $CaCo_3$	165	365	390	236
Influent magnesium, mg/l	14	34	—	—
Influent turbidity	10–60	Wells	Wells	Wells
Lime dosage, mg/l	90	227	250	150–160
Centrifuge, size	40 × 60 in.	40 × 60 in.	40 × 60 in.	40 × 60 in.
No of centrifuges	1	3	1	2
Solids concentrations, % solids				
From clarifier	5–10	2–4	12–18	12
From thickener	None used	15–25	20	40
From centrifuge	55–60	60	60	65
In centrate	0.5	5.5	3	5–10
Centrifuge loading				
gpd/machine	—	70,000	60,000	45,000
tons dry solids/day/machine	—	53	49	75
Capital costs, $[a]	80,000	1,750,000[b]	1,000,000[b]	1,650,000[b]
Annual costs, $	27,600	636,000	246,000	412,000
Total cost[c]				
$/mg	28.70	19	8	12
$/ton dry solids	28.70	8	10	10

[a]Estimated 1968 replacement costs.
[b]Includes recalcining plant.
[c]Exclusive of any income from sale of lime.

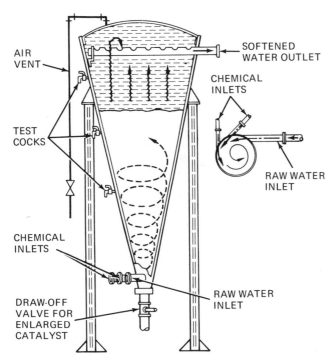

Fig. 5-8 Schematic diagram of a reactor for lime sludge pelletization. *(Courtesy Permutit Corp.)*

Lime Sludge Pelletization

Kneale[14] and Doe[3] have reported on a technique to provide a means of lime sludge disposal for plants in which recalcining or dewatering is not feasible nor attractive. In this approach, the water and the chemicals are introduced into a cone-shaped, bottom apex vertical tank tangentially at the apex (Figure 5-8 and 5-9). Sand (about 100 mesh in size) which serves as a nucleus for formation of calcium carbonate pellets is introduced at the top of the tank. The resulting sand–sludge mixture is maintained in suspension by the upward spiraling flow of the water. Calcium carbonate plates out on the sand, forming hard pellets (Figure 5-10) which grow in size until the vertical flow of water can no longer maintain the suspension. The pellets drop to the apex of the tank and are periodically discharged to a sump for dewatering. The pelletized sludge contains only 5–10 percent water. The limitations on this approach are: magnesium content should be less than 85 mg/l as $CaCO_3$; turbidity should be less than 10; in cold climates, the reactors must be enclosed in heated structures. Excessive magnesium forms magnesium hydroxide which does not plate out on

Fig. 5-9   Photo of sludge pelletization unit. *(Courtesy Permutit Corp.)*

the nuclei and will quickly clog downstream filters, Also, upflow rates (about 10 gpm/ft²) are too high to permit removal of suspended solids which will also pass on to down-stream filters. However, the resulting economics in sludge handling may be great enough to consider adding the reactor ahead of a conventional clarifier. The resulting pellets may be disposed of in landfill. Also, they have been used as abrasives for sandblasting, used as a soil conditioner, and used as a decorative gravel around shrubbery.

Pressure Filtration

The pressures which may be applied to a sludge for removal of water by the filter presses now available range from 5,000 to 20,000 times the force of gravity. In comparison, a centrifuge may provide

Fig. 5-10   Comparison of lime pellets and lime mud. *(Courtesy Permutit Corp.)*

forces of 3500 g and a vacuum filter, 1000 g. As a result of these greater pressures, filter presses may offer the following advantages:[15]

1. Filtration efficiency is improved, especially on materials which prove difficult to filter.
2. Requirements for chemical pretreatment are frequently reduced.
3. Solids concentration in the final cake is increased.
4. Filter cakes are more easily accommodated on material handling system.
5. Filtrate quality as measured by suspended solids content is improved.
6. Maintenance is minimal because very few moving parts are involved.

A filter press consists of a number of plates or trays which are held rigidly in a frame to ensure alignment and are pressed together either electromechanically or hydraulically, between a fixed and moving end. Figure 5-11 illustrates a typical filter press.

On the face of each individual plate is mounted a filter cloth. The sludge is fed by a suitable pumping method into the press and passes through feed holes in the trays along the length of the press until the cavities or chambers between the trays are completely filled. The sludge is then entrapped and a cake begins to form on the surface of the cloth. The water passes through the fibres of the cloth, and the solids are retained (Figure 5-12).

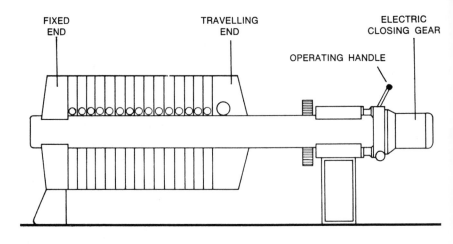

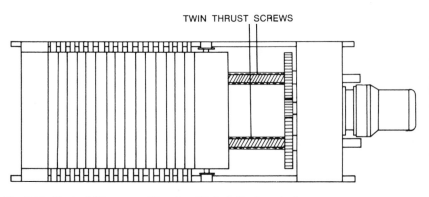

Fig. 5-11   Typical filter press. (*Courtesy Nichols Engineering*)

This process is repeated in each individual chamber until after a certain period — known as the "pressing cycle," which varies according to the particular type of sludge — the press is completely filled.

At the bottom of the tray drainage ports are provided. The filtrate is collected in these, taken to the end of the press and discharged to a common drain. At the commencement of the pressing cycle, the drainage from a large press can be in the order of 2 to 3000 gal per hr. This falls back rapidly to about 500 gal per hr as the cake forms and as the

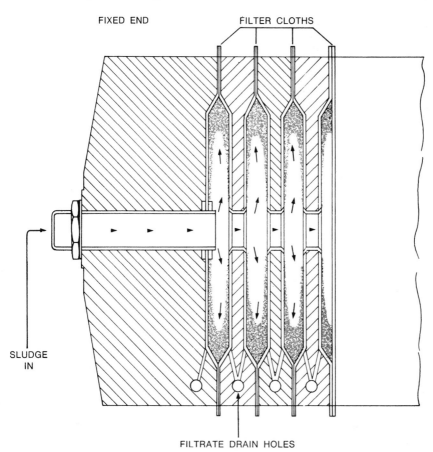

FIXED END   FILTER CLOTHS

SLUDGE
IN

FILTRATE DRAIN HOLES

Fig. 5-12   Feed flow and filtrate drainage. *(Courtesy Nichols Engineering)*

cake is full made up drops away virtually to nothing. The filter cakes
are then completely formed and ready for discharging.

Discharge of cakes from the filter press is accomplished as follows:

The pump feeding sludge to the press is stopped and any back pressure in the pipe work released through a bypass valve. The electrical
closing gear is then operated to draw the moving end back to its fully
retracted position. The individual press plates can then be moved in
turn over the gap between the plates and the moving end, thus allowing the filter cakes to fall out. This plate moving can be under manual
or fully automatic control. When all the plates have been moved and
the cakes released, the complete pack of plates is then pushed back
by the moving end and closed up under pressure from the electrical
closing gear. The valve to the press is then opened, the sludge feed

pump started and the next pressing cycle commences. Filter presses are normally installed well above floor level so that the cakes can drop out into trailers positioned underneath the press. Alternatively, conveyers can be installed under the presses to transport the cakes to the storage area. The dry cake may also be released from the filter plate by introducing compressed air behind the filter cloth on both sides of the plate, causing the cloth to flex, dislodging the cake. Although such a technique removes most of the cake, observations of operating presses indicate some manual attention is still required to remove all of the cake.

The basic requirements for a filter press system include (see Figure 5-13):

- Storage and mixing tanks for chemical reagents.
- A storage and conditioning tank, to provide a consistent feed to the filter presses for the duration of the filtration cycle. In this tank chemical reagents to improve the filterability characteristics of the sludge are introduced. Means for agitation are provided to prevent segregation of particles and also to prevent size degradation and breakdown in the flocculated feed.
- Feed pumps.
- Filter presses with ancillaries, including filter cloths and filtrate collection trays or launders. The selection of the correct filter cloth is an important factor in filtration; the choice for a particular duty is based on the chemical compatibility of the material from which the cloth is made with the material to be filtered, its retention of the particles present in the feed and the filter cake release from the cloth when the filter is opened.
- Interconnecting pipework, valves and air vessels for the feed system and filtrate disposal. Air vessels serve to even out pressure variations caused by the action of the feed pumps.
- Filter cloth washing machines and drying racks.
- Means for collecting and delivering filter cake to a disposal point in a form that is acceptable for any furthur processing required.

Although not used extensively in the United States, filter pressing of alum sludges has been used in Europe. It has been reported[2] that a plant in England dewaters a 1.8 percent alum sludge after conditioning with lime to a cake containing 25 percent solids with an 8 hr pressing time. Another plant found that pressing an untreated alum sludge with 1.5 to 2.0 percent solids yielded a wet cake containing 15 to 20 percent solids. With proper pretreatment with lime or polymer, it has been reported that cakes as dry as 30 to 50 percent solids can be produced from alum sludges. Lime sludges have been reported to dewater to 65 percent solids in only about 1 hr of pressing time.

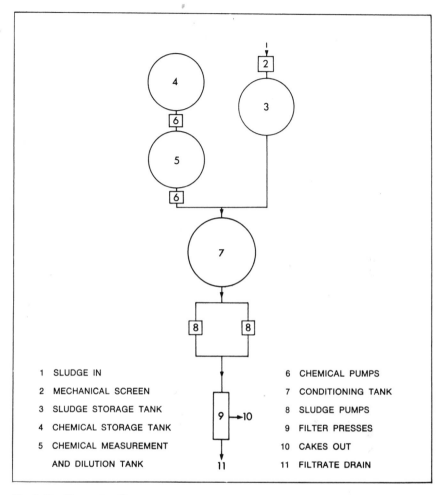

Fig. 5-13 Illustrative filter press system.

1	SLUDGE IN	6	CHEMICAL PUMPS
2	MECHANICAL SCREEN	7	CONDITIONING TANK
3	SLUDGE STORAGE TANK	8	SLUDGE PUMPS
4	CHEMICAL STORAGE TANK	9	FILTER PRESSES
5	CHEMICAL MEASUREMENT	10	CAKES OUT
	AND DILUTION TANK	11	FILTRATE DRAIN

Two disadvantages associated with past operations have been the short life of filter cloths and lack of automation. New, synthetic fibers have largely offset the first of these and has increased cloth life up to 12–18 months. Units with automated feeding and pressing cycles are also now available and it is only at the end of the cycle that the process becomes semiautomatic as no foolproof method of discharging the filter cake automatically is yet available. With proper selection of the filter cloth and careful control over the operating conditions, it is normally possible to achieve good filter cake discharge. A recent paper[22] summarizes the experiences in England at one plant with filter pressing of alum sludges. A 23 hr pressing time was found to increase the solids content of a polymer conditioned sludge from 3.5–6.25 percent to 22–25 percent while a 10–12 hr pressing time produced a 20 percent solids content. Filter fabrics of continuous filament polypropylene, monofilament courlene, nylon, and "Neotex 1188" (staple and filament nylon yarns with no backing) were tested with the Neotex 1188 proving most satisfactory. Cloth maintenance, exercised at 6 month intervals, consisted of washing the cloth in cold water, drying, and then brushing and shaking the cloth to remove fines. Four to five days of downtime were required for cloth cleaning and plate scraping. After 9 months, the cloth was in good shape and the pressed sludge still stripped well. The costs for the filter pressing system with a capacity of 4 tons/day of dry solids (193,000 gal of sludge per day at 5 percent solids) were about $180,000 including the presses; building, sludge conveyors, electrical, and pumps. Labor requirements were about 2 man hr per day for the 23 hr pressing cycle. The operating costs were as follows:

	Cost per Ton of Dry Solids
Electricity	$0.27
Fuel	1.30
Labor	3.50
Replacement materials	3.35
	$8.42

Including capital amortization charges, the total cost of pressing was estimated as $20.20/ton of dry solids. The estimated total cost of sludge freezing at the same plant was $55.50/ton.

Extensive tests of filter pressing of alum sludge have been conducted at Atlanta, Georgia[26] and have served as the basis of a full-scale filter press installation (55,000 lb/day dry solids). Attempts to filter the alum sludge without any conditioning agents proved impractical, confirming the need for proper conditioning. Fly ash proved to be an excellent conditioner. At a fly ash alum ratio of 1:1 (1 lb ash per 1 lb dry solids the sludge) a cake of 60 percent solids was achieved with a 165 min filter time with a filter pressure of 225 psi. Fly ash was used as a precoat. The use of lime as a conditioning agent also provided excellent results. The addition of quicklime in the amount of 12 percent by weight of the dry solids in the sludge produce a cake with 39.5 percent solids with a filter time of 165 min. Hydrated lime added in the range of 10–15 percent of the solids in the sludge provided filter cakes of 26–31 percent with pressing times of 75–105 min. The use of anionic polymers in amounts up to an equivalent polymer cost of $3/ton of dry solids did not produce good cakes. Pulverized clay was also tested as a conditioning aid but was found to be ineffective in that it formed a thin, blinding film on the filter medium. These tests at Atlanta indicate that with proper precoating and conditioning, filter pressing is capable of providing excellent dewatering of alum sludges.

Initial operating results from the full-scale installation at Atlanta are consistent with the above pilot results. Cakes of 40–50 percent solids have been produced using a 10 percent lime dosage and precoating.[29] The filters are 44 chamber machines (110 cubic ft) that operate at 225 psig with a total area of 1813 ft^2. The design cycle is 90 min which provides a gross loading of 316 ft^2/cycle at anticipated solids of 5 percent. A diatomaceous precoat of 100 lb/cycle is used. Typical filtrate suspended solids of less than 10 mg/l are reported. The March 1969 bid price for the filter press plant was $2,807,560.

European experience also indicates that pressure filtration of alum sludges is technically feasible. Experience at full-scale installations in the United States such as that at Atlanta will permit a better evaluation of costs in this country.

An extensive pilot test comparing filtering pressing and other alternate dewatering schemes was conducted at the Sobrante Filter Plant, part of the East Bay Municipal system in California.[30] Table 5-6 summarizes this comparative study.

Sobrante Filter Plant has a present nominal rated capacity of 40 mgd (60 mgd overload capacity) and provisions have been made for future expansion to an ultimate nominal capacity of 160 mgd. The Sobrante Filter Plant receives its raw water supply from San Pablo Reservoir, a 13.3 billion gallon capacity raw water storage reservoir situated in the hills along the easterly side of the San Francisco Bay Area. Approximately 85 percent of the water stored in San Pablo Reservoir is from the High Sierra, Mokelumne River watershed above Pardee Reservoir near Jackson, California. The water is conveyed to San Pablo Reservoir via a system of closed aqueducts. The remaining 15 percent of storage is local runoff from the reservoir's watershed. This watershed is mainly oak woodland with residential subdivisions and light commercial activities along the main tributary to the reservoir. The water taken from the reservoir is generally of high quality, with turbidities normally in the range of 2 to 10 JTU, except during winter periods of high local rainfall and runoff when the turbidity may go up to 150 JTU. This water is treated with liquid alum with conventional coagulation, flocculation, sedimentation, and rapid sand filtration. The water is chlorinated and adjusted to calcium carbonate saturation with unslaked lime after filtration prior to storage and distribution.

Sludge conditioning methods such as heat, pressure, and lime treatment were investigated in the field but rejected as ineffective or too expensive. Field tests confirmed that the preferred non-ionic dewatering polymer aids are the most economical; anionic polymers were

TABLE 5-6   ECONOMIC ANALYSIS — MACHINES AND OPERATING PARAMETERS[30]

Sobrante Filter Plant

Parameters	Scroll Centrifuge	Pressure Filter	Basket Centrifuge	Vacuum Filter
Solids loadings (dry basis)[a]				
Nominal–per unit				
Total	9,500	34,000	9,500	10,000
Overload–per unit	19,000	19,000	19,000	19,000
Total	34,000	68,000	22,000	20,000
(Overload conditions = 60,000 lb/day)				
Machine requirement				
Total	68,000	68,000	66,000	60,000
Machine size	28 gpm	64 × 81-in. chambers	120 gal bowl	416 ft.²
Assumed equipment life–years	20	30	20	25
Number for nominal load	2	½	2	2
Number for overload	2	1	3	3
Solids output				
Nominal–solids by wt., %	16	30	15	15
–gal/day of wet cake	12,700	10,600	13,500	13,500
Overload–solids by wt., %	16	20	10	10
–gal/day of wet cake	45,400	35,300	73,700	73,700
Polyelectrolyte dose (lb/T)				
Nominal load	3.0	4.0	4.5	1.5
Overload	8.0	4.0	5.0	2.0
Initial machine costs				
Machine cost–each	$40,000	$400,000	$ 45,000	$ 38,000
Total for overload	80,000	400,000	135,000	104,000
Annual cost (for 1,700 T/yr)				
Machine capital recovery @ 5%	$ 6,420	$ 26,020	$ 10,830	$ 7,380
Operation and maint–nominal	4,800	4,000	1,500	3,000
Power–nominal	3,850	1,190	3,210	10,900
Polyelectrolyte cost–nom (1,200 T/yr)	6,840	9,100	10,300	3,420
–overload (500 T/yr)	7,600	3,800	4,750	1,900
Disposal @ $5/cu yd[b]	56,100	34,400	68,700	68,700
Labor–½ man day/day (incl. benefits and overhead)	9,450	9,450	9,450	9,450
Total Annual Cost	$95,060	$ 87,960	$108,740	$104,750
Unit Cost $/T	56	52	64	62

[a]All solids loadings units are lb/day.
[b]Uncompacted unit weight in truck haul assumed = 70 lbs/ft².
T = Ton of dry solids

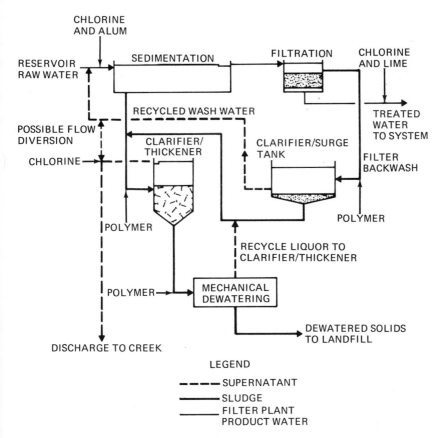

Fig. 5-14 Schematic diagram of the Sobrante, California sludge system.

somewhat effective and cationic polymers showed little or no effectiveness for this particular sludge.

Based upon the data summarized in Table 5-6 and considerations discussed below, a solid bowl scroll centrifuge was selected as the most flexible equipment for this application. The system shown in Figure 5-14 was scheduled for completion in early 1974 at an estimated cost of $2,450,000.

There are currently two options open for disposal of the sludge at Sobrante. The sludge can be thickened with a scroll centrifuge to 16 percent to 20 percent solids and hauled in an open dump truck to a sanitary landfill or, it can be thickened to approximately 12 percent solids with a reduced polymer dose using the scroll centrifuge and transported in a tank truck to a barge where the sludge would be mixed with thickened sewage sludge from the district's digestors and transported to the Sacramento-San Joaquin Delta area for land disposal.

The economic analysis demonstrated that the pressure filter is the most economical thickening scheme if sludge is hauled directly to a landfill from the mechanical dehydrator. However since a liquid sludge may be required for mixing with sewage sludge, the scroll centrifuge method was selected as being the most flexible since it provides a wide range of solids concentrations.

A total sludge drying system is being evaluated for use in conjunction with the scroll centrifuge. The combination of the scroll centrifuge with a pulse-jet drying system, based on preliminary studies, indicates that the total system will be more economical than a pressure filter alone. A 90 percent dry solids sludge appears to be possible with the drying system which dries sludge at temperatures in the range of 180° to 240°F.

Heat Treatment

When colloidal gel systems are heated under pressure, the gelatinous flocs break down, allowing bound water to escape which improves the dewaterability of the sludge. Sludges are pumped through a heat exchanger to a reaction vessel where it is held at elevated temperatures of 350–390°F at pressures of 180–210 psi when dealing with sewage sludges. Following about 30 min in the reaction vessel, the conditioned sludge passes back through the heat exchanger, gives up its heat to the incoming sludge, and enters a decanting vessel where the sludge settles to the bottom and is withdrawn for final dewatering.

It has been reported[3] that considerable improvement in alum sludge dewaterability can be realized at temperature and pressures below those normally used for sewage sludges. Testing has indicated that by heat treating alum sludges, filtration rates in the range of 10–20 lb dry solids/ft²/hr were not uncommon. Filter cake moistures averaged 79 percent.[27] Although heat treatment appears to offer an alternate means of alum sludge conditioning, there were little data available at the time of this writing.

Freezing

The process of freezing a sludge removes water from a gelatinous floc such as alum hydroxide sludge. When the sludge is thawed, the sludge does not revert to its original state but remains as small granular particles resembling brown sand with the volume reduced by a factor of 6. Thickening of the sludge prior to freezing is required to reduce the volume of material to be frozen to a minimum to reduce the costs. Freeze thawing a 2 percent alum sludge with separation of the sludge by gravity typically yields a sludge with about 20 percent solids which can be increased to 30–35 percent solids when filtered.

The freezing process has been used in batch-type plants treating alum sludge in England. A refrigeration plant provides the energy for both freezing and thawing the sludge in batch tanks. Ammonia refrigerant passes through coils in the batch tank causing freezing when the cold gas is circulated and thawing when the coils are used as condensers. By using multiple tanks, the process can be relatively continuous as the cold, vaporized ammonia passes through one set of tanks for freezing while the warm refrigerant passes through another set of tanks for thawing.

The Flyde plant in England uses the freezing process for water plant sludge. Operating costs of $5.00 to $9.50 per 1000 gal of sludge treated and capital costs of $17,000 per 1000 gal of sludge to be frozen per day have been reported.[2] Power requirements vary from 180 to 230 kwhr/1000 gal of sludge. A recent study[16] indicates that freezing rates as high as 40–60 mm/hr could be used to dewater sludge on a plant scale, higher than rates previously reported. This fact coupled with the fact that sludge may be frozen in thin layers on a flat surface without the need for compressive freezing gives some hope that a continuous, mechanized freezing process may be developed to overcome the present objections to operating and maintenance costs. The observations noted in the above study also led the authors to conclude that natural freezing of sludge in lagoons in cold regions could be made more effective by spraying sludge from the lagoon through a nozzle back onto the lagoon surface to melt snow cover, cool the sludge to freezing temperature as it travels through the cold air, and distribute sludge on a thin layer on already frozen sludge where it would freeze more readily. Costs of the freezing process have been estimated as $55.50 per ton of dry solids at one plant in England.[22]

COAGULANT RECOVERY

Alum Recovery

By acidifying aluminum hydroxide floc with sulfuric acid, it is possible to recover the alum as shown in the following equation:

$$2Al(OH)_3 + 3H_2SO_4 \rightarrow Al_2(SO_4)_3 + 6H_2O$$

By lowering the pH of the sludge to 2, it is possible to recover essentially all of the alum.[18] As indicated by the above equation, about 1.9 lb of sulfuric acid is required for each lb of sludge treated. Laboratory techniques for evaluating the feasibility of alum recovery for a given wastewater have been presented elsewhere.[18]

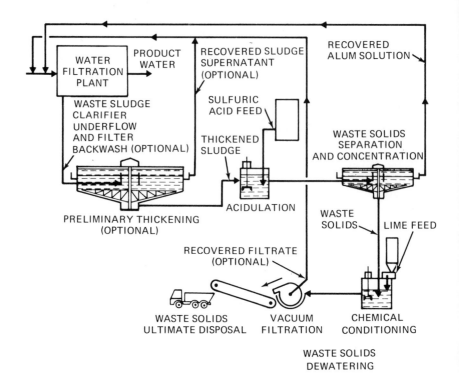

Fig. 5-15   Alum recovery flow sheet.[4]

Figure 5-15 presents a potential layout for an alum recovery system.[17] The waste sludge from the treatment plant may be thickened prior to the injection of the sulfuric acid. Following acidification, the waste solids are separated from the recovered alum in a gravity settling step. By recovering the alum, the solids requiring ultimate disposal are significantly reduced and the remaining solids can be dewatered for ultimate disposal. Numerous plants in Japan have used acid alum recovery with vacuum filtration of the waste solids to produce a cake of about 30 percent solids, although instances of the sludge being sticky and difficult to dewater have been reported. Solids-liquid separation in the acidification sludge can be improved by the use of the proper polymer. Use of 20–40 mg/l of cationic polymers have been reported to substantially improve this separation.[23] Alum recovery of about 60 percent is reported in many of these cases in Japan. Filter pressing has been reported to produce a 40–45 percent solids cake following acidification with 80–93 percent alum recovery achieved.[3] The Japanese experience indicates that optimum plant-scale recovery is obtained by using a pH of 2 by feeding about 1.2 times the stoichiometric requirements of sulfuric acid. The waste solids are reported to be about 15 percent solids prior to final dewatering. If iron and manganese are present in the alum sludge, they will also be recovered along with the alum, necessitating the blow down of some of the recovered alum to keep the iron and manganese content of the finished water in a satisfactory range. With a total loss of 25 percent alum in the blowdown and the loss with solids withdrawn from the system, iron and manganese would be concentrated by a factor of 3 in the raw water.

Experience with plant scale recovery of alum in Tampa, Florida[19] in the 1950s indicated that acid alum recovery permitted a savings of about \$12 per million gallons of water treated. However, recent investigators[3] of alum recovery express concern that the process may not provide a savings in terms of alum cost due to increased costs for sulfuric acid. Some savings in solids disposal will obviously result as the volume of solids requiring disposal is reduced by about a factor of 6. The full-scale Tampa alum recovery process was operated only for a short time[3] before being abandoned for several reasons. Among these reasons were the fact that a high color residual resulted which could not be eliminated except by high doses of chlorine, the constant, skilled operator attention required, the rapid changes in water quality, and the occasional use of softening processes which eliminated the production of aluminum hydroxide sludges.

In waters high in suspended solids, the amount of alum lost from the system will increase because of the greater volume of waste solids stream sent to the dewatering step.

A recent laboratory evaluation of the acid alum recovery process showed that the acidification of the alum sludge decreased the specific resistance (increasing the sludge's filterability) of the sludge from $3-5 \times 10^8$ sec^2/g to $1.5-2.6 \times 10^7$ sec^2/g when comparing lime-polymer treated alum sludges and cationic-polymer treated acidified sludges. Cake solids of 28–36 percent were obtained following acidification while solids of 16–25 percent were obtained with the conditioned alum sludge.

Although acid alum recovery is technically feasible, its economics must be evaluated in light of the circumstances faced for each specific potential application.

Magnesium Carbonate Coagulation and Recovery

Recent articles[20,21] presented details of a new coagulant recovery technique based on a combination of water softening and conventional coagulation techniques which can be applied to all types of waters. Magnesium carbonate is used as the coagulant with lime added to precipitate magnesium hydroxide as the active coagulant. The resulting sludge is composed of $CaCO_3 \cdot Mg(OH)_2$ and the turbidity removed from the raw water. The sludge is carbonated by injecting CO_2 gas which selectively dissolves the $Mg(OH)_2$. The carbonated sludge is then filtered with the magnesium being recovered as soluble magnesium bicarbonate in the filtrate which is recycled to the point of addition of chemicals to the raw water, reprecipitated as $Mg(OH)_2$, and a new treatment cycle begun. The resulting solids requiring disposal are the filter cake composed of $CaCO_3$ and the raw water turbidity. In large plants, it is possible to reduce the waste solids even further by slurrying the filter cake and separating the raw water turbidity from the $CaCO_3$ in a flotation cell. The purified $CaCO_3$ can then be dewatered and recalcined as high-quality quicklime. The stack gases from the recalcining operation then provide the CO_2 gas for carbonating the sludge for magnesium recovery and to recarbonate the water in the treatment plant. When this latter lime recovery step is practiced, the waste solids are reduced to only those which constituted the raw water turbidity.

Many waters contain enough magnesium that 100 percent magnesium recovery is achievable. With soft surface waters, magnesium recovery may be limited to about 80 percent.

Black et al.[20] has described a technique by which 20 major American cities softening hard, turbid waters could produce about 150,000 tons of $MgCO_3 \cdot 3H_2O$ for use as a coagulant by other plants in the process described above. In the process of doing so, the problem of disposal of sludges from the softening plants would be resolved. As noted earlier in this chapter, the lime softening of water with high magnesium concentrations results in objectionably large quantities of magnesium hydroxide appearing in the calcium carbonate sludges. Prior to 1957, it was considered impossible to recalcine the sludge produced by softening high-magnesium waters but it was then demonstrated that it was possible to dissolve selectively the $Mg(OH)_2$ from the $CaCO_3$ by carbonation with CO_2 gas. A plant using this concept has been in use in Dayton, Ohio since 1957. It is then possible to recover the magnesium from the carbonated sludge as shown in the following equations:

Sludge Carbonation: $Mg(OH)_2 + 2CO_2 \rightarrow Mg(HCO_3)_2$

Product Recovery: $Mg(HCO_3)_2 + 2H_2O \xrightarrow[\text{air}]{35-45°C} MgCO_3 \cdot 3H_2O + CO_2$

The magnesium bicarbonate solution clarified by either settling or filtration passes to a heat exchange unit where it is warmed to 35–45°C after which it is aerated by compressed air in a mechanically mixed basin. The $MgCO_3 \cdot 3H_2O$ precipitates rapidly (about 90 min) and the resulting product $MgCO_3 \cdot 3H_2O$ is vacuum filtered, dried, and bagged for shipment, hopefully to other plants where it may be used as a coagulant as described in more detail below. The calcium carbonate sludge remaining after the sludge carbonation step is dewatered and recalcined to produce a high quality quicklime.

The use of magnesium carbonate as a coagulant offers the potential for eliminating the problem of alum sludge disposal by replacing the alum sludges with a readily recoverable sludge. The chemical equations describing the coagulation process resulting from the recycled magnesium bicarbonate (as described at the first of this section) are as follows:

$$Mg(HCO_3)_2 + Ca(OH)_2 \rightarrow MgCO_3 + CaCO_3 + 2H_2O$$
$$MgCO_3 + Ca(OH)_2 \rightarrow Mg(OH)_2 + CaCO_3$$

The magnesium carbonate hydrolyzed with lime is as effective as alum for the removal of organic color and turbidity in many waters. The flocs formed are large and settle rapidly. When treating soft surface waters, this approach produces soft waters with sufficient calcium alkalinity to reduce corrosion. High carbonate hardness waters are softened by the process.

The lime doses required for the removal of free CO_2 and carbonate hardness are stoichiometric but the amount to be added for the precipitation of magnesium depends on a number of factors. This may be illustrated by the following example offered by Black et al.[21] A water containing both organic color and turbidity, has an alkalinity of 40 ppm and contains 30 ppm magnesium as $CaCO_3$. Jar tests show that the minimum effective amount of $Mg(OH)_2$ to be precipitated will be provided by a dosage of 60 mg/l $MgCO_3 \cdot 3H_2O$. The most economical treatment of this water requires that the entire dosage of 60 mg/l $MgCO_3 \cdot 3H_2O$ be added and that very little of the 30 ppm of magnesium present in the water be used. This results from four reasons:

(a) The dosage of lime required for the precipitation of the required amount of $Mg(OH)_2$ from 90 mg/l of $MgCO_3 \cdot 3H_2O$ is significantly less than that required to precipitate all of the $Mg(OH)_2$ from the 30 mg of $MgCO_3$ present plus 30 mg/l added, and the pH is significantly lower.

(b) The amount of CO_2 required to redissolve the $Mg(OH)_2$ from the sludge will be the same in either case, but, the amount of CO_2 required to recarbonate the high pH water of the second procedure will be much greater than the amount required for the lower pH water of the first procedure.

(c) After a few cycles a pH value is identified at which complete recovery and recycling of coagulant is achieved, after which no new coagulant is added.

(d) The unused 30 ppm of magnesium remaining in the water represents reserve coagulant which can be used if necessary by increased lime addition.

Although this example, using a water containing a significant amount of magnesium, is not typical of surface waters as a class, there are a number of cities, including Chicago, Detroit, Cleveland, Philadelphia, Pittsburgh, Indianapolis, Washington, D.C., and many smaller sized cities treating such waters. They should be able to substantially reduce their chemical treatment cost and at the same time eliminate or greatly reduce the extent of their sludge problems by adopting this technology.

TABLE 5-7 REQUIRED MAGNESIUM DOSE AS RELATED TO PHYSICAL AND CHEMICAL CHARACTERISTICS FOR 17 NATURAL WATERS[21]

Mg Required[a] (mg/l as $MgCO_3 \cdot 3H_2O$)	*Turbidity (JC)*	*Color (PrCo)*	*Alkalinity (mg/l)*	*Hardness (mg/l)*
40	165	50	13	16
30	105	30	17	17
10	4	26	4	5
15	10	12	74	83
22	13	4	54	84
10	14	10	17	17
25	106	11	51	110
24	50	15	41	71
30	24	30	27	43
35	7.5	27	10	12
35	6	5	92	127
25	2.5	0	80	100
24	41	14	34	69
32	104	38	11	13
21	2	4	12	40
20	7.5	8	71	86
22	15	24	48	71

[a]Precipitated as $Mg(OH)_2$.

Data on jar tests of magnesium carbonate coagulation on several natural waters have been presented.[21] For example, magnesium carbonate doses of 50 mg/l in conjunction with 125 mg/l of $Ca(OH)_2$ provided essentially the same degree of color and turbidity removal from one water as did a dose of 20 mg/l alum for a soft surface water. The magnesium carbonate technique often provides sufficient alkalinity in the finished water so that it can be effectively stabilized by pH adjustment where alum treatment may produce a water of such low alkalinity that it cannot be completely stabilized. The reduction or elimination of corrosion problems may be a significant point in many cases. Table 5-7 summarizes the required magnesium doses observed for 17 natural waters. A linear regression analysis made on the data

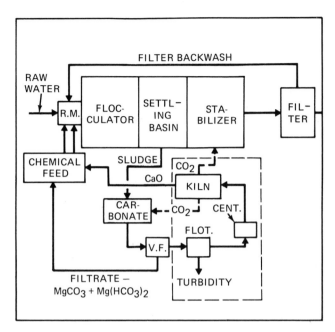

Fig. 5-16 Flow diagram of magnesium carbonate coagulation system.[21]

shown in Table 5-7 yielded the following equation for estimating the required magnesium dosage: Minimum magnesium dosage (mg/l of $MgCO_3 \cdot 3H_2O$) = 8.33 + 0.03 × turbidity + 0.46 × organic color − 0.03 × total alkalinity + 0.14 × total hardness.

The economics of applying the magnesium carbonate technology have been estimated for 17 major cities[21] and compared to the cost of coagulation for present treatment techniques using alum (and lime in some cases). If lime recovery is practiced so that CO_2 is provided at no cost, 13 of the 17 cities show a reduction in coagulant costs. If lime is purchased at $20/ton and CO_2 is obtained at no cost from a source in or near the plant (a 50 hp natural gas engine provides sufficient CO_2 for a one mgd plant), then 10 of the 17 cities would still show a reduction in cost. However, if both lime and CO_2 must both be purchased at a cost of $20/ton each, then only 2 of the 17 show a reduction in coagulant cost. These comparisons reflect only the cost of coagulants and do not reflect any credits related to the cost of sludge disposal. Figure 5-16 illustrates a potential flow sheet for the process. The sludge is carbonated in a basin with a detention time of about 1 hr to completely solubilize the magnesium with the separation from the solids being achieved by a vacuum filter. If the lime is not to be recovered, the resulting filter cake can be handled easily and disposed of in land fill or as an agricultural aid as a pH stabilizer for soil. If lime recovery is practiced, then the processes shown within the dashed lines are added. As discussed earlier, the calcium carbonate can be separated from the raw water turbidity in a flotation cell. With this approach, only the flotation unit skimmings require disposal and they are readily handled as they are primarily the raw water solids. The purified calcium carbonate sludge is then dewatered and calcined producing CaO and CO_2 needed in the process. For a more thorough evaluation of the sludge handling aspects, overall economic considerations, operation procedures, and instrumentation, a 2 year demonstration project is underway at Montgomery, Alabama using first a 50 gpm pilot plant and then the 20 mgd municipal plant.

REFERENCES

1. Dean, R. B., "Disposal of Wastes from Filter Plants and Coagulation Basins," *Journal American Water Works Association*, **45**, p. 1226 (Nov., 1953).
2. Krasauskas, J. W., "Review of Sludge Disposal Practices," *Journal American Water Works Association*, **61**, p. 225 (May, 1969).
3. "Disposal of Wastes From Water Treatment Plants," *Journal American Water Works Association*, *(August, 1969)*.
4. Fulton, G. P., "Disposal of Wastewater From Water Filtration Plants." *Journal American Water Works Association*, **61**, p. 322 (July, 1969).
5. Streicher, L., "Filter Backwash Gets Special Treatment," *The American City*, p. 94 (Nov., 1967).
6. Sutherland, E. R., "Treatment Plant Waste Disposal in Virginia," *Journal American Water Works Association*, **61**, p. 186 (April, 1969).
7. Neubauer, W. K., "Waste Alum Sludge Treatment," *Journal American Water Works Association*, **60**, p. 819 (July, 1968).
8. Farrell, J. B., Smith, J. E., Dean, R. B., Grossman, E., and Grant, O. L., "Natural Freezing for Dewatering of Aluminum Hydroxide Sludges," *Journal American Water Works Association*, **62**, p. 787 (December, 1970).
9. "Environmental Control Application Report," #ECE-001, Bird Machine Company (1971).
10. White, W. J., "A Centrifugal For Industrial Wastes," *Chemical Engineering Progress (June, 1969)*.
11. *Bird Machine Company Product Manual*, 1971.
12. Townsend, J. R., "What the Wastewater Plant Engineer Should Know About Centrifuges," *Water and Wastes Engineering* (Nov., Dec., 1969).
13. Super-O-Canter Continuous Solid Bowl Centrifuges, Bulletin 1287A, Sharples Stokes Division, Rennwalt Corporation, 1970.
14. Kneale, J. S., "Reduction of Sludge Volume From Lime Softening Plants," presented at the Central Section, Indiana American Water Works Association Meeting, September 16, 1971.
15. Thomas, C. M., "The Use of Filter Presses for the Dewatering of Sewage and Waste Treatment Sludges," presented at the 42nd conference of the Water Pollution Control Federation, Dallas, Texas, Oct., 1969.

16. Logsdon, G. S., and Edgerley, E., "Sludge Dewatering by Freezing," *Journal American Water Works Association,* **63,** p. 734 (Nov., 1971).
17. Albrecht, A. E., "Disposal of Alum Sludges," *Journal American Water Works Association,* **64,** p. 46 (January, 1972).
18. Culp, R. L., and Culp, G. L., *Advanced Wastewater Treatment,* Van Nostrand Reinhold, New York, 1971.
19. Roberts, J. M., and Roddy, C. P., "Recovery and Reuse of Alum Sludge at Tampa," *Journal American Water Works Association,* **52,** p. 857 (July, 1960).
20. Black, A. P., Shuey, B. S., and Fleming, P. J., "Recovery of Calcium and Magnesium Values," *Journal American Water Works Association,* **63,** p. 616 (October, 1971).
21. Thompson, C. G., Singley, J. E., and Black, A. P., "Magnesium Carbonate — A Recycled Coagulant," *Journal American Water Works Association,* Part I, **64,** p. 11, **64,** p. 93 (January and February, 1972).
22. Benn, D., and Bridge, L., 'Sludge Disposal by Pressing," *Journal of the Institution of Water Engineers,* **25,** p. 417 (November, 1971).
23. King, P. H., Bugg, H. M., Randall, C. W. "Improved Alum Recovery by Polymer Conditioning," unpuplished paper (1971).
24. Fulton, G. P., "Alum Recovery for Filtration Plant Waste Treatment," *Water and Wastes Engineering,* p. 78 (June, 1970).
25. Gates, C. D., and McDermott, R. F., "Characterization and Conditioning of Water Treatment Plant Sludge," *Journal American Water Works Association,* **60,** p. 33 (March, 1968).
26. "Test Results on Basin Sediment from the Chattahoochee Water Treatment Plant, City of Atlanta, Georgia," Beloit Passavant Corporation.
27. Schroeder, R. P., "Alum Sludge Disposal — Report No. 1," R & D Project No. DP-6551, Eimco Corporation, (Sept., 1970).
28. Mahoney, P. F., and Duensing, W. J., "Precoat Vacuum Filtration and Natural-Freeze Dewatering of Alum Sludge," *Journal American Water Works Association,* **64,** p. 665 (October, 1972).
29. Weir, P., "Research Activities by Water Utilities — Atlanta Water Dept.," *Journal American Water Works Association,* **64,** p. 634 (October, 1972).
30. Nielson, H. L., Carns, K. E., and DeBoice, J. N., "Scroll Centrifuge for Dewatering Wins at One Plant," *Water and Wastes Engineering,* p. 44 (March, 1973).

6 Disinfection

SIGNIFICANCE OF WATER BORNE DISEASES

It was not until the establishment of the germ theory of disease by Pasteur in the mid 1880s that water as a carrier of disease producing organisms really could be understood, although the epidemiological relation between water and disease had been suggested as early as 1854. In that year London experienced the "Broad Street Well" cholera epidemic and Dr. John Snow conducted his now famous epidemiological study. He concluded the well had become contaminated by a visitor who arrived in the vicinity with the disease. Cholera was one of the first diseases to be recognized as capable of being waterborne. Also, it probably is the first reported disease epidemic due to direct recycling of nondisinfected wastewater. Now, over 100 years later, the list of potential waterborne diseases due to microorganisms is considerably larger, and includes the bacterial, viral, and parasitic microorganisms shown below:[1]

Bacteria
  V. cholera
  Salmonella
  Shigella
Viruses
  Infectious hepatitis
  Coxsackie A and B (32 types)

Reoviruses (3 types)
ECHO viruses (34 types)
Adenoviruses (32 types)
Viral gastroenteritis
Viral diarrhea
Parasitic
  E. histolytica

Waterborne outbreaks of disease due to *V. cholera, Salmonella* and *Shigella* are well-documented. In addition to the Broad Street Pump cholera epidemic, the Hamburg-Altona Germany epidemic of 1892 is of interest. Itinerants with cholera were camped upstream of the two towns and polluted the river. Altona filtered its water from the river and developed no cases of cholera; Hamburg did not treat its water and the city developed a severe epidemic of cholera. Two recent California outbreaks of waterborne disease serve as sufficient evidence that *Salmonella* and *Shigella* can still cause problems in untreated water contaminated by wastes.

With respect to the viruses listed above, it is only with the virus(es) of infectious hepatitis that proven, documented outbreaks of waterborne virus disease have occurred. The December, 1955 epidemic in Delhi, India was the largest scale epidemic of hepatitis related to direct contamination of a water supply by sewage. However, since all of these viral agents are excreted in the feces of infected persons, they all are potential candidates for initiating waterborne epidemics. Particularly important are the rather elusive and ill-defined viruses which reportedly cause viral gastroenteritis and viral diarrhea. There have been reported 142 waterborne epidemics of these diseases in this country from 1946 to 1960 which affected more than 18,000 persons. Polio virus has frequently been mentioned as being possibly waterborne. Several epidemics of poliomyelitis attributable to water have been cited, but evidence has not been sufficiently complete to carry conviction. In the late 1930s it was discovered that the virus of poliomyelitis could be found in the feces of not only persons with paralytic disease, but also in the feces of asymptomatic carriers, and that the virus could be excreted in the feces of such individuals for several weeks. This observation led to the successful attempts to isolate poliovirus from sewage and from raw surface waters, and the hypothesis that waterborne transmission of poliomyelitis could occur. Epidemiological searches for episodes of waterborne poliomyelitis continued throughout much of the 1940s and the 1950s, and yielded 8 outbreaks in which contaminated drinking water was suspect, 1 of which occurred in the United States.

Waterborne outbreaks of disease caused by *E. Histolytica* are well-documented. The most infamous one in this country occurred in 1933 in Chicago. The cause proved to be plumbing inadequacies which allowed fecal pollution of the drinking water of the Congress and Ambassador Hotels, resulting in many cases of amoebic dysentery and some 23 deaths.

Coxsackie viruses have long been recognized as the causative agents of idiopathic myocarditis and recently the echoviruses have also been attached to this list. The Coxsackie viruses are known to have a high affinity for heart tissues. When such infection occurs in the uterus, the

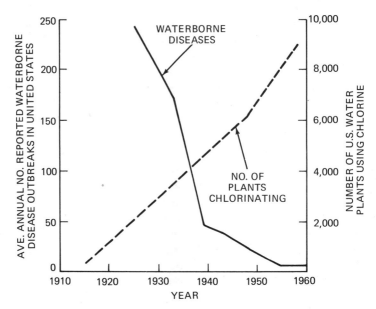

Fig. 6-1 Average annual number of waterborne disease outbreaks 1920–1960 and the number of water plants using chlorine.

infant may develop a congenital heart anomaly. The incidence of this particular disease in the United States is estimated to be approximately 12,000 cases per year.

To illustrate the current trend of human enteric diseases, the total annual incidence of typhoid fever has dropped off continuously from approximately 2,000 in 1952 to 300 in 1968. The annual hepatitis incidence, however, has increased or at least remained at a level of 50,000 to 60,000. It is generally agreed that probably one out of ten hepatitis infections results in clinical illness. On this basis, the number of infections resulting from this virus is approximately 500,000 per year in the United States.

The number of disease outbreaks identified as waterborne in the United States has dropped steadily following the introduction of chlorination of potable water. Figure 6-1 shows the annual number of reported waterborne disease outbreaks from 1920 through 1960 and the growth of U. S. water treatment plants using chlorine disinfection from 1910 through 1958. There is an obvious relationship between the drop in waterborne disease outbreaks and the number of water treatment plants using chlorine disinfection. Of course, improvements in other water treatment procedures — flocculation, filtration, etc., have contributed to the decrease in waterborne disease, but disinfection is, and will remain, the most important treatment process for the prevention of waterborne disease.

Although the decline in waterborne disease outbreaks over the last 60 years is marked, there has been no decline since the mid-fifties and indeed, there has been a slight increase (see Figure 6-2). Two particularly large outbreaks occurred in 1961–1965 in community systems (Riverside, California and Gainesville, Florida) with the highest number of cases per outbreak since 1938. Table 6-1 presents a summary of the known outbreaks of waterborne diseases for the 10 year period 1961–70.

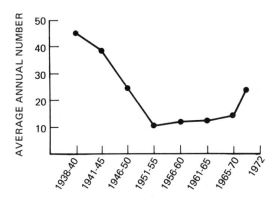

Fig. 6-2 Average annual number of waterborne outbreaks 1938–1972.

Two agents never before associated with documented waterborne outbreaks in the U. S. appeared during the 1961–70 period: enteropathogenic E. coli (EEC) and Giardia lamblia. Various serotypes of EEC have been implicated as the etiologic agent responsible for disease in newborn infants, usually the result of cross contamination in nurseries. Now, there are several well-documented outbreaks of EEC (serotypes 0111:B4 and 0124:B27) associated with adult waterborne disease.

Giardia lamblia is a flagellated protozoan responsible for giardiasis. The three outbreaks of giardiasis occurred in resort or recreational areas. Another protozoal infection sometimes transmitted by the waterborne route is amoebiasis or amoebic dysentery, which occurred in 3 outbreaks. The etiologic agent for this disease is Entamoeba histolytica, an amoeba.

The major cause of outbreaks in public systems is through contamination of the distribution system — primarily via cross connections and back siphonage. However, when cases of illness are considered, the picture changes. Outbreaks resulting from contamination of the distribution system are usually quite contained and few illnesses result. However, contamination of the source or a breakdown in treatment, produces many illnesses. When cases of illness are considered, the major perpetrators are source contamination and treatment deficiencies.[6] About 46 percent of the outbreaks in public systems were found to have been related to these two particular causes; *92 percent of the cases of illness in the public systems occurred as the result of source and treatment deficiencies.*

The Riverside, California salmonella outbreak in 1965 was the largest of its kind with an estimated 16,000 cases. At least 70 were hospitalized and 3 deaths occurred. The supply was a groundwater supply that had not been chlorinated and initiation of chlorination stopped the outbreak. Just why the outbreak occurred when it did is not known. The water supply was clearly incriminated but the exact source of contamination could not be determined. Of interest is the fact that before the outbreak no coliform organisms were detected during the routine surveillance of the water system. During the investigation, S. *typhimurium* was isolated from the water. A surprising result of the quantitative analysis of a composite water sample from all parts of the city was that *Salmonella* were estimated to be 10 times as numerous as *E. coli.*

The fall of 1972 saw the popular press featuring a story on the isolation of viruses from the drinking water of Bellerica and Lawrence, Massachusetts. Both towns were reportedly served by water treatment systems of high quality. Although the water treatment profession can point with pride to the major contribution it has made to the public health of the United States over the last 60 years, the Riverside and Massachussetts instances illustrate that complacency is certainly out of order and continuing efforts to better understand disinfection and to universally apply (chlorination was *not* even in use at Riverside) proven technology are essential.

DISINFECTION BY COAGULATION, SETTLING, FILTRATION AND ADSORPTION PROCESSES

Although the addition of halogens, such as chlorine, is immediately called to mind when one considers disinfection, other unit processes also provide rather substantial degrees of disinfection.

The chemical coagulation process has been noted by several researchers as providing high degrees of removal of viruses. For example, alum coagulation was found to remove 95 to 99 percent of Coxsackie virus and ferric chloride was found to remove 92 to 94 percent of the same virus.[7] With both alum and $FeCl_3$, good virus removal was contingent upon good floc formation and the absence of interfering substances. The effectiveness of removal was not temperature-dependent with either coagulant. It has been postulated that the

TABLE 6-1 WATERBORNE DISEASE OUTBREAKS 1961–70 BY TYPE OF ILLNESS AND SYSTEM

Illness	Private Systems		Public Systems		Total	
	Outbreaks	Cases	Outbreaks	Cases	Outbreaks	Cases
Gastroenteritis	25	4,498	14	22,048	39	26,546
Infectious hepatitis	22	664	8[a]	239	30[a]	903
Shigellosis	16	939	3	727	19	1,666
Typhoid	14	104			14	104
Salmonellosis	4[b]	96	5	16,610	9+	16,706
Chemical poisoning	7	42	2	4	9	46
Enteropathogenic E. coli	4	188			4	188
Giardiasis	1	19	2	157	3	176
Amoebiasis	2	14	1	25	3	39
Total	95	6,564	35	39,810	130	46,374

[a]One gastroenteritis outbreak also included seven cases of infectious hepatitis.
[b]One gastroenteritis outbreak was preceded by outbreak of 38 cases of salmonellosis.

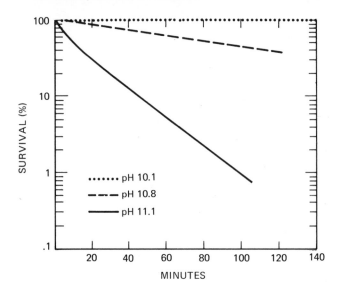

Fig. 6-3 Inactivation of Poliovirus 1 by high pH at 25°C in lime-treated (500 mg/l Ca (OH)₂), sand-filtered secondary effluents.

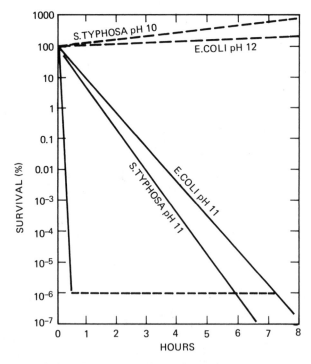

Fig. 6-4 Survival of *E. coli* at pH 11 and 12 and *S. typhosa* at pH 10, 11, and 12 at 25°C.

aluminum coordinates with the carboxyl groups in the virus's protein coat, followed by incorporation of the complex into the precipitating hydrated aluminum oxide. It has been determined that viruses were not inactivated by alum coagulation but could be partially recovered from the sludge. The presence of organic material was shown to decrease the amount of virus removed by alum or ferric chloride coagulation.

Lime coagulation has demonstrated the ability to effectively remove and inactivate viruses. The mechanism of inactivation under alkaline conditions is probably caused by denaturation of the protein coat and by disruption of the virus. In some cases, complete loss of structural integrity of the virus may occur under high pH conditions.

The pH reached is a critical factor in determining the degree of virus inactivation. Figure 6-3 shows the marked difference in virus inactivation as the pH is increased from 10.1 to 10.8 and then to 11.1. These data indicate that it may be useful to increase the pH to slightly higher levels during excess lime softening than is now done so as to obtain higher inactivations during softening. Figure 6-4 illustrates the effects of pH on *E. Coli* and *S. typhosa*.

Virus removal by filtration has also been studied.[16] Up to 98 percent of poliovirus Type I was removed in a coal and sand filter at filtration rates from 2 to 6 gpm/ft^2, when a low dose of alum was fed ahead of the filters. Up to 99 percent was removed when the alum dose was increased and conventional flocculation and sedimentation carried out ahead of the filter.

Floc breakthrough of the filter, sufficient to cause a turbidity of less than 0.5 units, was usually accompanied by virus breakthrough. Addition of a polyelectrolyte to the filter influent decreased virus in the effluent by increasing floc strength and floc retention in the filter. The subsequent chapter on wastewater in water supplies further discusses the roles of coagulation and filtration in removal of viruses.

While the bulk of studies on virus removal have been laboratory studies, Walton[9] reported a study of coliform bacteria removal in over 80 full-scale water treatment plants. He found that the removal of coliform bacteria by coagulation, sedimentation, and filtration averaged about 98 percent for the several plants studied. The average coliform densities in filtered but unchlorinated waters ranged from 2.9 to 200 per 100 ml. These data make it clear that chlorination must also be provided if adequate coliform removals are always to be achieved.

Although the coagulation and filtration processes cannot be relied upon to provide the only means of pathogen removal, these processes perform an important role in assuring the ability to completely disinfect the water. Hudson[17] noted the importance of minimizing finished water turbidity to maximize assurance of disinfection.

He concluded that filtration plants operated to attain a high degree of removal of one impurity tend to accomplish high removals of other suspended materials. Examples of parallelism in removal of turbidity, manganese, microorganisms, and bacteria were observed. Plants producing very clear water also tended to secure low bacterial counts accompanied by low incidence of viral diseases. He also concluded that the production of high-quality water requires striving toward high goals as measured by several, not just one or two, quality criteria. These criteria include filtered water turbidity, bacteria as indicated by plate counts and by presumptive and confirmed coliform determinations, and thorough chlorination. The operating data he reviewed for plants treating polluted water indicate that low virus disease rates occur in cities where the water treatment operators aim to produce a superior product rather than a tolerable water.

The report of an AWWA Committee on viruses in water[18] noted that viruses, because of their small size, more easily become enmeshed in a protective coating of turbidity-contributing matter than bacteria would. For most effective disinfection they concluded, turbidities should be kept below 1 Jackson Unit; indeed, they felt it would be best to keep the turbidity as low as 0.1 Unit, as recommended by AWWA water quality goals. They noted that the limit of 5 Jackson Units of turbidity specified in the USPHS "Drinking Water Standards," 1962, is meant to apply to protected watersheds and not to filtration plant effluents. With turbidities as low as 0.1 to 1, they concluded that a preplant chlorine feed need be only enough to have a 1 mg/l free chlorine residual after 30 min contact time. Postchlorination practice would then depend upon the ability to maintain such residuals throughout the distribution system.

Thus, the disinfection of water is dependent upon optimizing the unit processes of chemical coagulation, sedimentation, and filtration (which in themselves provide substantial reductions in bacteria and viruses) to produce minimum water turbidities to assure the maximum contact between any remaining pathogens and the disinfectant added. The interdependence of these unit processes and chlorination to assure adequate disinfection are graphically illustrated by the reported results[27] of a study in 1945, in Philadelphia, Pa. Members of Civilian Public Service Unit No. 140 served as experimental subjects in studies conducted by representatives of the University of Pennsylvania and the armed forces. These volunteers consumed water innoculated with known quantities of coliforms and infectious hepatitis virus. Some of this contaminated water was subjected to various water treatment processes. Hepatitis syndrome developed in one or more volunteers who consumed the water in each type of treatment except in those cases where adequate chlorination inactivated the virus. However, treatment other than chlorination reduced the incidence of infection by about 40 percent.

The report on this experiment concluded with the following:

(a) Coagulation settling and filtration (diatomite filter) of contaminated water did not eliminate or inactivate infectious hepatitis virus . . .

(b) The application to such water (previously coagulated, settled and filtered) of sufficient chlorine to provide, after 30 minutes contact, total and free residual chlorine concentrations of 1.1 and 0.4 ppm, respectively, apparently was adequate to inactivate the hepatitis virus under the conditions of this experiment. However, the same 30 minute residual total chlorine concentration (1 part per million) in contaminated water that had not been pretreated by coagulation, settling and filtration, did not inactivate the infectious hepatitis virus.

CHLORINATION

History

Chlorine was first formed over 200 years ago from the union of manganese dioxide and hydrochloric acid by the Swedish chemist Karl Scheele. It was not until 1810 that this element was given the name, chlorine (from the Greek work chloros meaning green), by an Englishman, Sir Humphry Davy, who established that Scheele had indeed discovered a chemical element.

According to the AWWA,[2] the earliest recorded use of chlorine directly for water disinfection was on an experimental basis, in connection with filtration studies at Louisville, Ky., in 1896. It was then employed for a short period in 1897 in England, again on an experimental basis, to sterilize water distribution mains following a typhoid epidemic. Its first continuous use was in Belgium, beginning in 1902, for the dual objective of aiding coagulation and making water biologically "safe." In North America, the first continuous, municipal application of chlorine to water was in 1908 — to disinfect the 40 mgd Boonton reservoir supply of the Jersey City, N. J., water utility.

Contested in the courts, the practice was declared to represent a public health safeguard, and this action paved the way for its rapid extension to other public water supplies throughout North America and elsewhere.

Growth of the chlorine industry was slow until the last decade of the nineteenth century. During this period electrolytic chlorine production became a commercial reality. Chlorine liquefaction also was developed and refined during this period. Commercial production of liquid chlorine in the United States began in 1909 with the filling and shipping of the first chlorine cylinder. This development opened new frontiers and paved the way for the great surge of the industry that followed over the next 50 years. Vehicular transfer of chlorine was supplemented the following year when the first 16 ton, single-unit railroad tank car of chlorine was shipped.

In 1910, Fort Meyer, Va., used liquid chlorine on an experimental basis for water disinfection by using simple dry gas feeders. Solution feeders were employed 2 years late at Niagara Falls, N. Y. to control a recurring outbreak of typhoid fever. Following the introduction of improved equipment for feeding of liquid chlorine, hypochlorite water chlorination gradually decreased in popularity, but it received renewed stimulus with the commercial availability, in 1928, of high-test calcium hypochlorite, a more stable and active material than the various bleaching powders and solutions previously available. Today, the principal sources of chlorine for water disinfection are: (1) in elemental form, as a liquefied compressed gas, (2) as calcium hypochlorite, (3) as chlorine bleach solutions (not to be confused with elemental liquid chlorine or with water solutions of chlorine gas), and (4) as chlorine dioxide. Although the bactericidal properties of chlorine dioxide are about equal to those of chlorine (when optimally applied), it rarely is used solely for the purpose of disinfection. The literature contains several reports, however, of its successful application for oxidation of iron, manganese, phenolic and chlorophenolic compounds, for control of certain algae, and for control of taste and odor problems.

Refinements in feed equipment and establishment of rational control procedures has led to nearly universal (but most unfortunately not yet universal) use of chlorination in the water treatment industry in the United States. In 1962, the latest year for which published data are available, more than 10,000 U. S. water treatment facilities, serving about 135 million persons, employed chlorine disinfection.

Alternate Forms of Chlorine Available

For purposes of disinfection of municipal supplies, chlorine is primarily used in two forms: as a gaseous element or as a solid or liquid chlorine-containing hypochlorite compound. Gaseous chlorine is generally considered the least costly form of chlorine that can be used in large facilities. Hypochlorite forms have been used primarily in small systems (less than 5000 persons) or in large systems where safety concerns related to handling the gaseous form outweigh economic concerns.

Gaseous Chlorine. In the gaseous state, chlorine is greenish yellow in color and about 2.48 times as heavy as air; the liquid is amber colored and about 1.44 times as heavy as water. Unconfined liquid chlorine rapidly vaporizes to gas; 1 volume of liquid yields about 450 volumes of gas.

Chlorine is only slightly soluble in water, its maximum solubility being approximately 1 percent at 49.2°F. Chlorine confined in a container may exist as a gas, as a liquid, or as a mixture of both. Thus, any consideration of liquid chlorine includes that of gaseous chlorine. Vapor pressure of chlorine in a container is a function of temperature, and is independent of the contained volume of chlorine; therefore, gage pressure is not an indication of container contents.

Chlorine gas is a respiratory irritant with concentrations in air of 3–5 ppm by volume being readily detectable. Chlorine can cause varying degrees of irritation of the skin, mucous membranes, and the respiratory system, depending on the concentration and duration of exposure. In extreme cases, death can occur from suffocation. The severely irritating effect of the gas makes it unlikely that any person will remain in a chlorine contaminated atmosphere unless he is unconscious or trapped. Liquid chlorine may cause skin and eye burns upon contact with these tissues; when unconfined in a container, it rapidly vaporizes to gas producing the above effects.

Chlorine is shipped in cylinders, tank cars, tank trucks, and barges as a liquefied gas under pressure. Containers commonly employed include 100 and 150 lb cylinders, 1 ton containers, 16, 30, 55, 85, and 90 ton tank cars, 15 to 16 ton tank trucks, and barges varying from 55 to 1110 ton capacity. Regulations limit the maximum filling of chlorine cylinders to 150 lb. Most cylinders are designed to withstand a test pressure of 480 psig. Ton containers, loaded to about 2000 lb of chlorine and having a gross weight as high as 3700 lb, are designed to withstand a test pressure of 500 psig. Single unit chlorine tank cars have 4 in. corkboard or self-extinguishing foam insulation and are designed to withstand pressures of either 300 or 500 psig. Chlorine tank barges have either 4 or 6 tanks, each with a capacity of from 85–185 tons.

Hypochlorites. Calcium hypochlorite, a dry bleach, has been used widely since the late 1920s. Present day commercial high test calcium hypochlorite products, contain at least 70 percent available chlorine and have from less than 3 percent to about 5 percent lime.

Calcium hypochlorite is readily soluble in water, varying from about 21.5 g/100 ml at 0°C, to 23.4 g/100 ml at 40°C. Tablet forms dissolve more slowly than the granular materials, and provide a fairly steady source of available chlorine over an 18–24 hr period.

Granular forms usually are shipped in 35 lb or 100 lb drums, cartons containing 3 3/4 lb resealable cans, or cases containing nine 5 lb resealable cans. Tablet forms are shipped in 35 lb and 100 lb drums, and in cases, containing twelve 3 3/4 lb or 4 lb resealable plastic containers.

Commercial sodium hypochlorite usually contains 12 to 15 percent available chlorine at the time of manufacture, and is available only in liquid form. It is marketed in carboys and rubber-lined drums of up to

TABLE 6-2   STABILITY OF NaOCl SOLUTIONS

Available Chlorine Trade percent	Chlorine, g/l	Half-Life Days, 77°F
3	30	1700
6	60	700
9	90	250
12	120	180
15	150	100
18	180	60

50 gal volume, and in trucks. All NaOCl solutions are unstable to some degree and deteriorate more rapidly than calcium hypochlorite (see Table 6-2). The effect can be minimized by care in the manufacturing processes and by controlling the alkalinity of the solution. Greatest stability is attained with a pH close to 11.0, and with the absence of heavy metal cations. Storage temperatures should not exceed about 85°F; above that level, the rate of decomposition rapidly increases. While storage in a cool, darkened area very greatly limits the deterioration rate, most large manufacturers recommend a maximum shelf life of 60–90 days. All hypochlorite solutions are corrosive to some degree and affect the skin, eyes, and other body tissues that they contact.

The costs of sodium hypochlorite are usually 20–25 cents/gal (1973 prices) which is equivalent to 13–19 cents/lb of available chlorine. This contrasts to chlorine gas costs of 3–5 cents/lb in large quantities, 6–7 cents/lb in ton containers, and 8–15 cents/lb in small cylinders. Table 6-3 summarizes the characteristics of typical sodium hypochlorite solutions.

On-site Manufacture of Hypochlorite

Chlorine gas, although economical, is risky to store and use. This is true especially in populated areas where the release of a 50 ton tank, for example, could require the evacuation of a 5 square mile area. If this were to occur in a highly populated area, it is evident that the results could be disasterous. In order to eliminate the risks inherent in chlorine storage and use, some large cities, such as New York, Chicago and Providence have shifted to the use of concentrated solutions of sodium hypochlorite.

Sodium hypochlorite is the form of chlorine that comes closest to being as economically feasible as liquid chlorine as a chlorinating agent. It is clearly a safer form to handle. The major reason sodium hypochlorite is not more widely used as a chlorinating agent is based

TABLE 6-3 CHARACTERISTICS OF TYPICAL SODIUM HYPOCHLORITE SOLUTIONS

	Formula	NaOCl
	Molecular Weight	74.45
	Strength, NaOCl	1 to 20% by vol
		1 to 16% by wt

	Household Grade	Industrial Grades
Available chlorine (% by vol)	5	15
Density, lbs/gal	9.0	10.05
Freezing or crystallization point	24.8°F	−1.4°F
Specific gravity 20°C	1.075	1.205
Viscosity, Saybolt sec. univ. 75°F	30.7	34.6
Available chlorine, g/l	50	150
lbs/gal	0.42	1.25
Free caustic as NaOH, g/l	1.0 min	4 to 15
Free carbonate as Na2CO3 g/l	0.5	2.5
Iron as Fe+++, mg/l	0.1 min	0.5 min
	0.5 max	1.0 max
pH	10.6	10.82 to 11.2
Color	Pale greenish yellow	Pale greenish yellow
Odor	Chlorinous	Chlorinous
Physical state	Liquid	Liquid
Stability	Decomposes in hot water. Sensitive to light, heat with slow decomposition.	

on economics. Any slight dosage savings do not offset the increased cost of sodium hypochlorite compared to liquid chlorine. Other hypochlorite salts, such as calcium hypochlorite, are several times more costly than sodium hypochlorite solutions.

Although the relative safety of sodium hypochlorite is well-known, the large quantities that must be used presents a problem. For example, a water treatment plant using 6 tons of chlorine per day requires a storage of 83 tons of hypochlorite per day because it is used in maximum concentrations of 15 percent. This requires transportation and storage costs of 30,000 tons per year. Transportation and storage costs can be seen as substantial factors in the cost of a liquid sodium hypochlorite operation. Large quantities cannot be economically stored because of the short half-life of 15 percent concentration.

A solution to the storage and safety problem is to use a system which can produce at the treatment plant sodium hypochlorite as needed. There have been recently introduced several commercial[37,38,39,40] systems which can produce hypochlorite electrolytically from sodium chloride. These systems are being evaluated in major installations. For example,[35] New York City is about to start testing three systems for on-site generation of sodium hypochlorite used to disinfect effluents from the city's sewage treatment plants.

New York now buys about 8 million gal/year of 15 percent hypochlorite solution for $1.3 million. The three electrolytic systems to be tested at New York will produce hypochlorite from sea water or brine. They are the Pepchlor unit made by Pacific Engineering and Production Co. of Nevada (Henderson, Nev.); the Sea-Cell unit, manufactured by A. Johnson & Co., Ltd. (London), and marketed by Parkson Corp. (Fort Lauderdale, Fla.); and Diamond Shamrock's Sanilec unit. New York pays 13 cents/lb for hypochlorite (16 cents/gal) of 15 percent solution. Estimates for on-site hypochlorite production in New York vary from 3 to 5.5 cents per lb.

The electrolytic cell converts the chloride ion to hypochlorite ion as follows:

$$NaCl + H_2O \xrightarrow{e} NaOCl + H_2 \uparrow$$

The raw materials required are salt (either in a brine solution or in seawater), power, and water. Since any calcium and magnesium present in the brine form precipitates which can foul the system, it is desirable that both the salt and water be as hardness-free as possible. Ordinary rock salt contains sufficient hardness that use of evaporated salt, solar salt, or southern rock salt is recommended. In installations where either softened, de-ionized water, or tap water of sufficient quality is not available, water softeners with automatic regeneration cycles should be included in the system.

One manufacturer's system is illustrated in Figure 6-5. Concentrated brine is automatically produced in a salt dissolver which is designed for pneumatic salt delivery by truck. Water, introduced through headers approximately 3 feet above the bottom of the tank, dissolves the salt while passing through the salt bed to collectors on the bottom. As the salt dissolves, dry salt shifts down to replace it. The brine is then transferred to a brine storage tank either by gravity or by a brine transfer pump. The water feed to the automatic brine makeup system and the liquid level in the brine storage tank are controlled by automatic liquid level controls located in each unit.

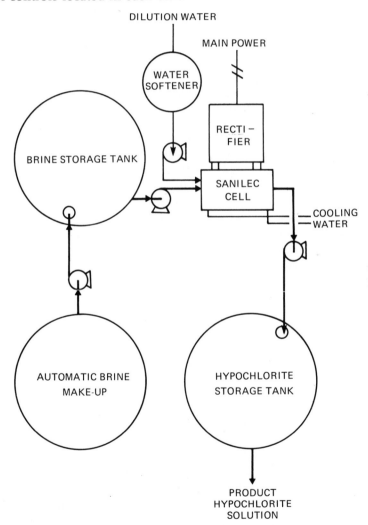

Fig. 6-5 Illustrative hypochlorite generation system. (*Courtesy Diamond Shamrock Co.*)

Concentrated brine is pumped by a metering pump to the cell inlet where it is diluted to the proper concentration (3 percent) with water and discharged into the cell when it is electrolyzed. Low level electrodes in the brine storage tank will alarm and stop the pump in the event of insufficient brine supply. Hypochlorite is generated continuously or intermittently, depending upon the chlorination practices at the point of use. The main power circuit and transfer pumps are automatically activated or deactivated by electrical signals from level controls in the hypochlorite storage tank. No special maintenance attention is required other than periodic replacement of the electrodes, routine pump inspection, and supervision during salt deliveries.

Safety features will alarm and shut down the cell to prevent possible malfunction. These are a low-flow alarm on the inlet water dilution line, a thermal switch in the cell, and a high-low voltage control for the cell. Corrosion of nearby equipment from stray electric currents is prevented by grounding the inlet of the cell and allowing the cell effluent to discharge through a breaker box.

Centrifugal pumps transfer the hypochlorite solution to a fiberglass holding tank with capacity for one day's supply. Upon signal from a chlorination cycle timer, an automatic valve on the tank outlet will open and discharge the hypochlorite by gravity into the intake of the condenser cooling water circulation pumps. The Diamond Shamrock system discussed above uses an electrode reported[35] to be dimensionally-stable titanium coated with ruthenium.

Other commercial systems use other electrodes. The Pepcon system[40] uses an anode consisting of an electroplated metal (lead) oxide coating on graphite or titanium. Advantages claimed for this anode are: (1) it is relatively hard (about 5 on the MOH scale) and thus resists abrasion, (2) its electrical resistivity is low (40–50×10^{-6} ohm-cm^{-1}) and thus is more conductive than mercury or graphite, (3) lead dioxide is insoluble in water and becomes essentially inert when held at its highest valence state by use as an anode, (4) it is readily available, (5) it is low in cost.

A typical system arrangement recommended by this manufacturer is shown in Figure 6-6. Table 6-4 presents the manufacturer's data on estimated power requirements.

Another manufacturer (Engelhard Industries) uses platinum-plated titanium anodes. The bulk of experience with this system has been with sea water as the source of salt. About 1 lb of equivalent chlorine is produced from each 120,000 gal of salt water processed. Anode life of 5–7 years is claimed. This system is typically manufactured in cell modules of 1/2 lb/hr or larger capacity. All materials are PVC or titanium and are corrosion resistant. The cells are arranged in series with as many as 20 cells in an electrical series. The system is usually operable from a central control panel and is designed for automatic

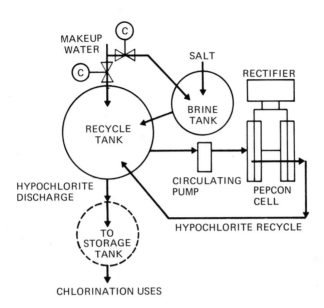

Fig. 6-6   Hypochlorite generation system. *(Courtesy Pacific Engineering and Production Co. of Nevada)*

operation. Automatic shutdown features are incorporated to secure the power in case of insufficient flow through any cell.

A different approach has been developed and is being marketed by Ionics.[34,39] In this approach, a membrane cell (Figure 6-7) is used. A stack of these cells is the heart of the overall system (which is shown in Figure 6-8).

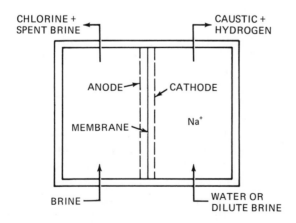

Fig. 6-7   Membrane cell used in Ionics system.

TABLE 6-4   ESTIMATED POWER REQUIREMENTS FOR HYPOCHLORITE PRODUCTION

Daily Sodium Hypochlorite Production (In lbs)		Kw hrs Required	
Using Salt	Using Clean Seawater	Using Salt	Using Clean Seawater
38	48	152	144
76	96	304	288
114	144	456	432
152	192	608	576
190	240	760	720
285	360	1140	1080
380	480	1520	1440
570	720	2280	2160
760	960	3040	2880
950	1200	3800	3600
1140	1440	4560	4320
1330	1680	5320	5040
1520	1920	6080	5760
1710	2160	6840	6480
1900	2400	7600	7200

Saturated brine is fed to the anode compartment. At the anode surface, chlorine gas is generated. The effluent from the anode compartment is sent to a gas-liquid separator where the chlorine gas is separated from the depleted brine, which is discarded. Hydrogen gas and hydroxyl ions are generated at the cathode. The effluent from this compartment goes to a gas-liquid separator in which the hydrogen is removed from the caustic solution. The hydrogen is diluted with air to a harmless level and vented to the atmosphere.

Chlorine and caustic are fed to the reactor, with the caustic in slight excess. The reactor is water cooled to avoid decomposition of hypochlorite and formation of oxygen. Insoluble gases, mainly oxygen, are removed in a gas-liquid separator before the hypochlorite is sent to the product storage tank.

Brine feed for the system comes from the salt storage tank (lixator) which is used for both salt storage and production of saturated brine. Because of the crystal properties of sodium chloride, the bed of salt acts as a filter for the saturated brine. In applications where the hypochlorite is not used as it is produced, a storage tank may be provided for it.

The anode is an expanded titanium screen coated with a special metal-oxide coating. Anode life of 3 years is projected.[34] The cathode is constructed from a sheet of expanded mild steel.

The membrane is a new ion exchange plastic membrane developed

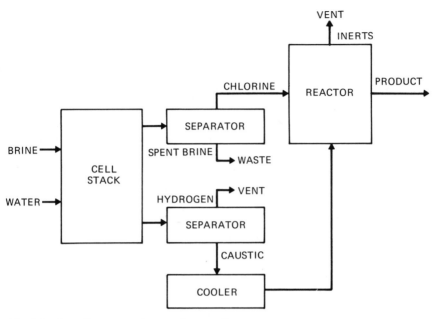

Fig. 6-8 Overall system using membrane cells.

by DuPont and is called "Nafion" per fluorosulfonic acid membrane. The membrane acts as a barrier in the cell allowing sodium ions to pass into the cathode compartment but excludes chlorine gas and brine. It also prevents hydroxyl ions from migrating into the anode compartment which would cause a loss in current efficiency by generation of oxygen gas. Membrane life of at least 2 years is projected.[34]

The cell typically produces hypochlorite with 8 percent available chlorine at 1.7 kw hr/lb available chlorine. Higher concentration of 12 percent can be generated at the expense of power — 2.1 kw hr/lb available chlorine.

TABLE 6-5   POWER AND SALT REQUIREMENTS FOR ON-SITE HYPOCHLORITE GENERATION AS REPORTED BY MANUFACTURERS

	Power kw hr/lb NaOCl	Salt lbs/lb NaOCl
Ionics "Cloromat"	1.7	2.1
Engelhard "Chloropac"	2.8	(Sea water)
Pepcon "Pep-Clor"	3.5	3.25
Diamond Shamrock "Sanilec"	2.5	3.05

Power and chemical requirements for the various systems based upon the manufacturer's data are compared in Table 6-5. At power costs of 1 cent per kw hr and salt costs of $10 per ton, the costs for power and salt are 2.7–5.1 cents per lb of sodium hypochlorite. To this, of course, must be added electrode replacement costs, operating labor and capital amortization. All manufacturers are confident that total costs will be substantially less than purchased hypochlorite even when sea water is not available and competitive with gaseous chlorine in smaller plants. For example, the Ionics system estimated costs[34] for 1000 lb/day production system are total costs of 5.4 cents (4.0 operating and 1.4 capital) per lb of available chlorine. Truck delivered hypochlorite costs 15 to 25 cents per lb of available chlorine and gaseous chlorine costs 5 to 8 cents per lb in ton cylinders. Tank car gaseous chlorine use would be clearly cheaper but the number of plants using tank car quantities is indeed low.

Chlorine Reactions

When chlorine is dissolved in water at temperatures between 49°F and 212°F, it reacts to form hypochlorous and hydrochloric acids:

$$Cl_2 + H_2O \rightarrow HOCl + HCl$$

This reaction is essentially complete within a very few seconds. The hypochlorous acid ionizes or dissociates practically instantaneously into hydrogen and hypochlorite ions: $HOCl \rightleftharpoons H^+ + OCl^-$. These reactions represent the basis for use of chlorine in most sanitary applications.

Hypochlorite chlorine forms also ionize in water and yield hypochlorite ion which establishes equilibrium with hydrogen ions:

$$CaOCl_2 + 2H_2O \rightarrow 2HOCl + Ca(OH)_2$$
$$NaOCl + H_2O \rightarrow HOCl + NaOH$$

Chlorine has a strong affinity for other materials, particularly reducing agents, which sometimes exert a considerable chlorine demand. In these reactions, the chlorine atom manifests its great tendency to lose electrons to form chloride ion or organic chlorides. Reacting substances typically include Fe^{++}, Mn^{++}, NO_2^-, H_2S and the greater part of organic material present. The reactions with inorganic reducing substances are generally very rapid; those with organic materials are generally slow, the extent depending on the kind and form of excess available chlorine present.

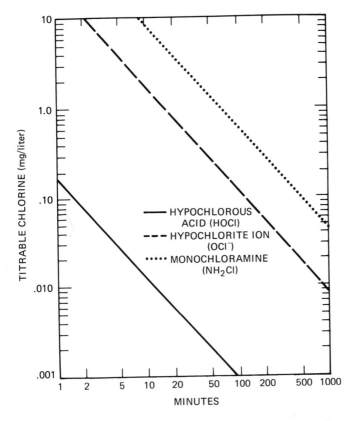

Fig. 6-9 Relationship between concentration and time for 99 percent destruction of *E. coli* by 3 forms of chlorine at 2–6°C.

Resistance of Pathogens to Chlorination

Obviously, it is important to know the relative resistance of virus, bacteria, and cysts to chlorine if one is to be assured that chlorination practice is going to be adequate. There are wide differences in the susceptibility of various pathogens to chlorine. It is now agreed that the waterborne pathogens in order of disinfection difficulty generally is bacteria < virus < cysts and that the difficulty increases in sewage or waters contaminated with sewage. The destruction of pathogens by chlorination is dependent upon water temperature, pH, time of contact, degree of mixing (as discussed later), turbidity, presence of interfering substances, and concentration of chlorine available. Chloramines formed if ammonia is present are much less effective disinfectants. For example, in one study,[29] at normal pH values, approximately 40 times more chloramine than free chlorine was required to produce a near 100

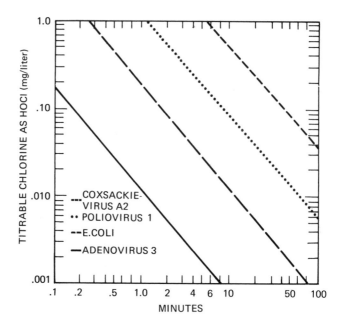

Fig. 6-10 Relationship between concentration and time for 99 percent destruction of *E. coli* and 3 viruses by hypochlorous acid (HOCl) at 0–6°C.

percent kill of *E. coli* in the same time period. For *S. typhosa*, this ratio was about 25 to 1. To obtain a near 100 percent kill with the same amounts of residual chloramine and free chlorine required approximately 100 times the contact period for chloramine.

Figure 6-9 illustrates the relationship between chlorine concentration and the contact time required for 99 percent destruction of *E. coli* for 3 different forms of chlorine. Figure 6-10 illustrates the relative resistance of three viruses and *E. coli*. While the polio and Coxsackie viruses are considerably more resistant than *E. coli* to HOCl, the adenovirus tested is apparently more sensitive. One-tenth ppm of chlorine as HOCl destroyed 99 percent of *E. coli* in about 99 sec. The same quantity of the adenovirus was destroyed in about one third of that time by the same amount of HOCl, but at this HOCl concentration the same amount of poliovirus required almost 85 min more and 5 times longer, and the Coxsackie virus required over 40 min, more than 24 times longer than the *E. coli*. Thus, the same equilibria are established in water regardless of whether elemental chlorine or hypochlorites are employed.

The reactions depend on the pH of the water as indicated in Figure 6-11 Chlorine exists predominately as HOCl at low pH levels. Between

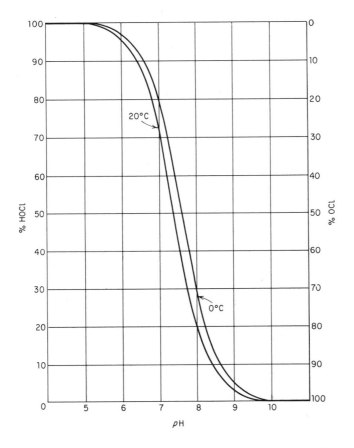

Fig. 6-11 Relative amounts of HOCl and OCl⁻ formed at various pH levels.

pH 6.0 and 8.5 there occurs a very sharp change from undissociated to completely dissociated hypochlorous acid. Above pH 7.5, hypochlorite ions predominate, while above pH 9.5 chlorine exists almost entirely as OCl⁻. The significant difference between elemental and hypochlorite chlorine forms is the resultant pH and the influence of same on the relative amounts of HOCl and OCl⁻ existing at equilibrium. Chlorine tends to decrease pH while hypochlorites tend to increase it.

Generally, the HOCl form has been considered to be a far more effective disinfectant than OCl⁻. Several investigators have reported that HOCl is 70–80 times as bactericidal as OCl⁻ and that increasing the pH reduces germicidal efficiency because most of the free chlorine exists as the less microbiocidal OCl⁻ at the higher pH levels. However, one recent study[43] reported that one virus (poliovirus 1) was more rapidly inactivated at pH levels (pH = 10) where the free chlorine is in the form of OCl^- rather than HOCl. This finding, contrary to the conclusions reached by many others which are presented in the next section, was undergoing evaluation on other animal viruses at the time of this writing.

The reactions of chlorine with ammonia in solution are of great significance in water and wastewater treatment. In the presence of ammonia or other nitrogenous material, a complex of chloramines is obtained:

$$NH_3 + HOCl \rightarrow NH_2Cl + H_2O \text{ (Monochloramine)}$$
$$NH_3 + 2HOCl \rightarrow NHCl\ NHCl_2 + 2H_2O \text{ (Dichloramine)}$$
$$NH_3 + 3HOCl \rightarrow NCl_3 + 3H_2O \text{ (Trichloramine)}$$

These chloramines have many different properties than HOCl and OCl^- forms. They exist in various proportions depending on the relative rates of formation of monochloramine and dichloramine which change with the relative concentrations of chlorine and ammonia, as well as with pH and temperature. Above pH about 9, monochloramines exist almost exclusively; at pH about 6.5, monochloramines and dichloramines coexist in approximately equal amounts; and below pH 6.5 dichloramines predominate, while trichloramines exist below pH about 4.5.

The point where all ammonia is converted to trichloramine or oxidized to free nitrogen is referred to as the breakpoint. Chlorination below this level is combined available residual chlorination; that above this level is free available residual chlorination.

Beginning about 1915, the influence of ammonia on the disinfecting capacity of chlorine was observed and reported by many investigators. The widespread adoption of chlorine-ammonia treatment followed recognition that the combination of chlorine and ammonia produced a more stable disinfecting residual than that produced by chlorine alone, and that the process could be applied to limit the development of objectionable tastes. Chlorine-ammonia treatment later declined in popularity, however, largely due to the realization of the superior bactericidal efficiency of hypochlorous acid. Currently, the principal application is for post-treatment, to provide long-lasting chloramine residuals in potable water distribution systems.

Just as do free-available chlorine forms differ in germicidal capacity, so do inorganic and organic chloramine combined-available chlorine forms. Inorganic chloramines have a substantially lower oxidation-reduction potential and less germicidal capacity than HOCl; most organic chloramines have little or no germicidal capacity and are of concern in water treatment practice primarily for the chlorine they con-

TABLE 6-6 VIRAL INACTIVATION RATE CONSTANTS AT VARYING pH OF SEWAGE AT 10°C

Sewage pH	Flash mixing of 3^ mg/l chlorine Viral Inactivation Rate Constant (min⁻¹)	Free Chlorine Duration (sec)	Viral Kill (percent)
5.0	120.0	110.0	>99.9999
5.5	72.0	24.0	>99.9999
6.0	44.0	10.0	99.93
6.5	26.0	5.6	92.0
7.0	16.0	5.4	76.0
7.5	9.5	5.0	55.0
8.0	5.6	5.0	40.0
8.5	3.5	4.9	25.0[a]
9.0	2.0	4.8	15.0[a]
9.5	1.2	4.6	10.0[a]

[a]Essentially control survival.

sume that otherwise would be available for disinfection, and in the measurement and interpretation of residual chlorine. Other researchers[29] conclude that with the exception of adenovirus 3, from 3 to 100 times more chlorine is required to kill the viruses than E. coli, A. aerogenes, E. typhosa, or Sh. dysenteriae.

Both pH and temperature have a marked effect on the rate of virus kill by chlorine. Table 6-6 summarizes data on virus inactivation rates at varying pH at 10°C in filtered secondary sewage effluent. The constants in Table 6-6 were determined from experimental results using model f_2 virus which was seeded in filtered secondary sewage effluent buffered to the desired pH value. Chlorine solution was flash mixed at a dose of 30 mg/l and the time of persistence of any free chlorine as well as virus survival was determined.[25]

Since most sewage effluents are near neutral pH, the 30 mg/l chlorine dose would result in less than 80 percent viral inactivation. However, by merely lowering the pH to 6.0, viral kill could be increased from 80 to 99.3 percent. The results dramatically show how the kill of virus may be enhanced, especially when hypochlorites are fed for sewage disinfection by proper pH control. For 4 logs (99.99 percent) of viral inactivation, the flash mixing of the following dosages were required: 40 mg/l of chlorine solution (pH 6.3), 50 mg/l of hypochlorite solution (pH 7.4) and 25 mg/l hypochlorite in acidified sewage (pH 5.0).

Other studies show that decreasing the pH from 7.0 to 6.0 reduced the required virus inactivation time by about 50 percent and that a rise in pH from 7.0 to 8.8 or 9.0 increased the inactivation period about 6 times. In the presence of low free chlorine residuals, the virus inactivation rate is markedly affected by variations of temperature and pH. In the destruction of viruses by chlorine, Clarke's work suggests that the temperature coefficient for a 10°C change is in the range of 2 to 3, indicating that the inactivation time must be increased 2 to 3 times when the temperature is lowered 10°C. Data also indicate that the chlorine concentration coefficient lies in the range of 0.7 to 0.9. This means that the inactivation time is reduced a little less than half when the free chlorine concentration is doubled. To increase virus kill, therefore, there is some advantage in increasing the contact time instead of raising the chlorine content. The observation has been confirmed in that one study found increased contact time to be more important than increased chlorine dosage in the inactivation of the T2 phage in excess of 90 percent.[23] Inactivation of poliovirus in water also has been found to be more dependent on contact time.

The effects of chlorine on a mixture of cysts consisting predominantly of Entamoeba histotyica and Entamoeba coli has been reported.[44] The cysticidal efficiency was related to the residual concentration of halogen measured chemically after 10 min of contact time. It was concluded that the chemical species of free and combined halogens which prevail in low pH waters were superior cysticides when compared to forms which predominate in waters of high pH. Cysts were not affected by the H+ concentration alone (pH 5 to pH 10); however, the possibility that a low pH enhanced the ability of the halogen to penetrate the cyst wall cannot be dismissed.

The most rapid acting cysticide was found to be the hypochlorous acid (HOCl) form of free available chlorine. Figure 6-12 presents the results. In reviewing Figure 6-12, it should be remembered that the effect of cysticidal halogen is influenced by the concentration of cysts present; the greater the cyst concentration, the greater the halogen dose required. Since polluted waters usually contain no more than 5 cysts/ml, the 1600–2000/ml employed in the study are probably greater than normally required. The results of this study have practical applications in water treatment facilities in that the evidence shows that cysts are not likely to survive where prechlorination to a free residual is practiced. As a result of the chemical reactions between the coagulants normally employed and natural alkalinity, the water remaining in the coagulation-settling basin is usually at a pH where most of the chlorine will be HOCl, the superior cysticide. Free chlorine residual remaining in the distribution system is not of equal cysticidal potency because of the common water works practice of raising the pH for corrosion control results in the OCl⁻ species of free available chlorine predominating.

The AWWA Committee report on virus in water[18] concluded that "in the prechlorination of raw water, any enteric virus so far studied would

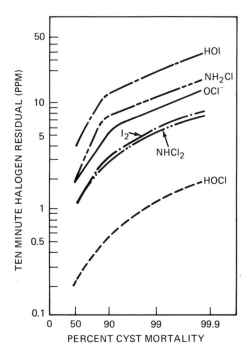

Fig. 6-12 Ten minute residual concentrations of iodine and chlorine species required to obtain the indicated levels of cyst mortality in ten minutes at 30°C. 1600–2000 amoebic cysts from simian hosts/ml.[44]

be destroyed by a free chlorine residual of about 1.0 ppm, provided this concentration could be maintained for about 30 min and that the virus was not embedded in particulate material. In postchlorination practices where relatively low chlorine residuals are usually maintained, and in water of about 20°C and pH values not more than 8.0–8.5, a free chlorine residual of 0.2–0.3 ppm would probably destroy in 30 min most viruses so far examined."

The Committee also states "there is no doubt that water can be treated so that it is always free from infectious microorganisms — it will be biologically safe. Adequate treatment means clarification (coagulation, sedimentation, and filtration), followed by effective disinfection. Effective disinfection can be carried out only on water free from suspended material." They caution that "positive coliform tests certainly indicate possible virus contamination, but a negative coliform test may not indicate freedom from viruses."

There is certainly a need for improved virus detection techniques in light of the greater susceptibility of the coliform indicator organism to destruction by chlorination. In the interim, application of the estab-

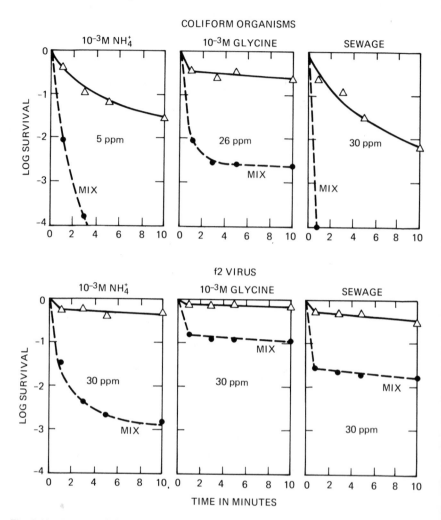

Fig. 6-13 Bactericidal and viricidal effect of flash mixing of chlorine in waters containing nitrogenous interferring substances: $10^{-3}M$ ammonium ion, $10^{-3}M$ glycine, and filtered secondary sewage effluent. All buffered to pH 7.0 and cooled to 0°C. Chlorine dosed as indicated at time zero to two test volumes, one flash mixed, the other slowly added with gentle stirring.

lished basis for minimizing the probability for viral survival (low turbidity, low pH, high free chlorine residual, adequate mixing, long contact times) must be relied upon.

The Importance of Mixing

In addition to the chemical and water quality considerations discussed above, a complete and uniform mixing of the chlorine with the water to be disinfected is important. Many full-scale chlorination facilities are inadequately designed with the chlorine applied directly to the chlorine contact basin, sometimes even without a diffuser. Flows in the contact basins are usually highly stratified, resulting in very little dispersion of the chlorine. Disinfection is one process for which any measurable short circuiting is completely ruinous to process efficiency. Attention to tank shape mixing and to proper baffling is critical. A laboratory study[25] graphically illustrates the importance of adequate mixing in assuring disinfection. The results are presented in Figure 6-13 which is self-explanatory.

Recent studies at the Sanitary Engineering Research Laboratory of the University of California evaluated the importance of rapid initial mixing in chlorination. The researchers[32] believed that prolonged segregation of the chlorine from the bacteria results in poorer performance. The bactericidal compounds are formed by the reaction of chlorine with nitrogenous and carbonaceous matter to form chlorine complexes. The speed of these reactions varies; some are quite fast while others are relatively slow. They concluded that the residuals initially formed are apparently much more bactericidal than the compounds formed later. They concluded that rapid initial mixing allows these more lethal residuals to come in contact with the bacteria and act as killing agents.[42]

Collins[32] also concluded that backmixing that occurs in a propeller-mixed tank may contribute to poorer disinfection. They noted that reactions may take place between the more bactericidal residuals initially formed upon entering the reactor and the more complex residuals which form over a longer time. Thus, these simple and more bactericidal compounds may be "destroyed" before they can act.

In one experiment, chlorination was effected in two alternative ways. In one arrangement, the chlorine was added directly to a constant flow stirred tank reactor (CSTR) (t = 37.5 min) near the paddle which rotated at 50 rpm. Under the alternative arrangement, the chlorine was added to a short (t = 0.12 min) tubular reactor placed upstream from the CSTR. The tubular reactor consisted of a 1/4 in. PVC tee acting as a constriction in a 3/4 in. line. The chlorine was added through the stem of the tee. Settled wastewater was the effluent to be treated and an aqueous solution of chlorine gas was the disinfecting agent.

Table 6-7 summarizes the data they obtained with these two systems. Although the total mean residence time was essentially the same, the kill obtained when the chlorine was added at the head of the tubular reactor was much greater. The coliform survival ratio — the parameter used to indicate bactericidal effectiveness — was about 1 order of magnitude lower in this case. However, the coliform survival ratio for the CSTR alone was approximately the same regardless of whether the chlorine was added directly to the CSTR or to the tubular reactor. The increase in kill was attributed solely to the more rapid mixing which occurred in the tubular reactor.

Another result is that for each run, the chlorine residual in the CSTR was slightly lower when the chlorine was applied to the tubular reactor. It was concluded that the various reactions involved were driven to a greater degree of completion because of the more rapid mixing.

Collins concluded that improved performance is provided by applying the chlorine to plug flow reactors designed to provide rapid initial mixing. Where mechanically stirred mixers are necessary, the chlorine solution should be applied to the stream ahead of the mixer. The two process streams (chlorine solution and water) would not be applied directly and separately to the mechanical mixer.

A subsequent study[42] was made on both a pilot scale and plant scale (2.4 mgd). To parallel the tubular reactor used by Collins, a diffusion grid was installed in the full scale plant in a 27 in. effluent pipe. The data from the plant scale study (which were consistent with the pilot scale tests) revealed that while rapid initial mixing provided improved performance, i.e., lower coliform survival ratios, at low values of Rt (R is the total chlorine residual, t is the contact time), a somewhat lesser degree of initial mixing produced the same results when t was increased with R maintained constant. Thus, in situations where a long pipeline is used to provide long contact time for disinfection, a reasonably good initial mixing device will probably produce the best obtainable results.

As pointed out by Stenquist and Kaufman,[42] the reason for the merging of results with varying degrees of mixing at the higher values of Rt is unclear. It may be related to varying resistance of the organisms to disinfection. The more resistant organisms, if not killed by the highly bactericidal residuals which are initially formed and which can be utilized under conditions of rapid initial mixing, must be subjected to the same value of Rt as would have been the case if the initial mixing had been relatively slow. Whatever the actual reason, the answer seems to lie in the mechanism of chlorine disinfection and the reactions which chlorine undergoes in wastewater.

The advantages of the high initial kill which rapid initial mixing provides are of significant importance. For example, existing plants which have CSTR reactors for chlorination could be improved by installing a chlorine injection device in the pipe, if one exists, leading to the chlorine contact unit. Use of tubular reactors also provides a much better residence time distribution than a CSTR. This is particularly important in chlorination because the coliform kill is so strongly time dependent. Use of a CSTR allows "short circuiting" of fluid particles with short chlorine contact times and thus a very high concentration of coliform organisms.

**TABLE 6-7   EFFECT OF METHOD OF APPLYING CHLORINE ON REDUC-
TION OF COLIFORM ORGANISMS**[42]

Run No.	Chlorine Residual mg/l	Coliform Survival Ratio
A. Chlorine Applied in Tubular Reactor (t = 0.12 min) Ahead of CSTR (t = 37.5 min)		
1	3.2	0.00286
2	5.1	0.00419
3	8.1	0.00049
B. Chlorine Applied Directly to CSTR		
1	4.2	0.0378
2	6.1	0.052
3	8.8	0.012

Chlorine Feed

The literature including the textbooks, is replete with descriptions of conventional chlorine feed systems and safety procedures
and a replay of this information would serve no useful purpose here.
However, the recent introduction of an all-vacuum system for feeding
chlorine (now offered by several manufacturers such as Capital Controls Co., Badger Meter Mfg. Co., Fischer and Porter) is of interest.[45]

The gas lines and the chlorine solution lines that have been under
pressure in previous equipment have been eliminated. With this development, the chlorinator may be mounted directly on the cylinder,
thus eliminating one of the major remaining problem spots in the
chlorine gas feeding system.

Comparison of the typical all-vacuum and the pressure-vacuum-
pressure systems reveals that the all-vacuum system eliminates all
manifold piping and flexible connectors as well as isolating auxiliary
valves, thus removing one of the major safety problems in chlorine gas
handling. In addition, the often troublesome chlorine solution line,
containing undissolved gas and of considerable length, now can be
removed, thus making the operation safer and the installation less expensive, using a vacuum line.

The requirement of indoor installations for chlorinators which previously was necessary to keep a certain amount of heat on the gas
pressure lines so that reliquification did not occur, now can be eliminated because gas is under vacuum on exit from the cylinder and will

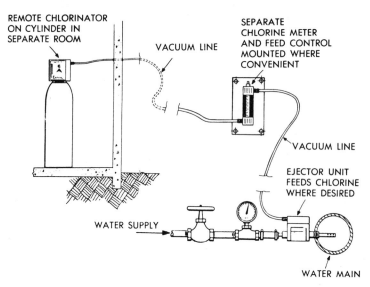

REMOTE CHLORINATOR
ON CYLINDER IN
SEPARATE ROOM

VACUUM LINE

SEPARATE
CHLORINE METER
AND FEED CONTROL
MOUNTED WHERE
CONVENIENT

VACUUM LINE

EJECTOR UNIT
FEEDS CHLORINE
WHERE DESIRED

WATER SUPPLY

WATER MAIN

Fig. 6-14   Typical vacuum chlorine feed system. *(Courtesy Capital Controls Co., Inc.)*

reliquify only if the temperature drops to approximately 40° to 50°F below zero (−40° to −45.5°C) — an unlikely operating condition.

Figure 6-14 illustrates a typical system using this equipment. Units can be mounted on ton cylinders as well. The operation of the system is relatively simple. Water under pressure passes through the ejector at high velocity causing a vacuum which opens a spring loaded diaphragm check valve in the ejector body. This vacuum is transmitted through a plastic line to the separate chlorine flow meter and flow rate adjusting valve. From here, the vacuum extends to the chlorinator through the other connecting line.

With vacuum at the chlorinator (about 2 lb below atmosphere) the main diaphragm moves backward opening a spring loaded valve at the inlet of the chlorinator. Chlorine at cylinder pressure then enters through the inlet valve where the pressure is reduced to a vacuum. This reduced pressure (vacuum) is controlled at a constant value regardless of changes of cylinder pressure. The inlet valve is forced closed if water flow at the ejector is stopped or if there is any opening in the vacuum system. Gas pressure in the system is prevented by an emergency vent valve in the chlorinator which would open and relieve the pressure to the outside through a vent line. Loss of chlorine supply is shown on the chlorinator by an off pressure gage.

Chlorine under vacuum is conducted to the separate meter unit where the flow rate is read directly in lb/24 hr (PPD). It then passes through the adjusting valve at the top of the meter where chlorine rate is set as desired. This unit also compensates for vacuum changes at the ejector caused by pressure changes of the ejector water supply or the injection back pressure. Chlorine then passes under vacuum to the ejector where it mixes with water and is injected as chlorine-water solution.

OTHER DISINFECTANTS

Iodine

Iodine possesses the highest atomic weight of the four halogens and is the least soluble in water with a solubility in water of 339mg/l (and the least hydrolyzed by water); it has the lowest oxidation potential, and reacts least readily with organic compounds. Taken collectively, the characteristics mean that low iodine residuals should be more stable and, therefore, persist longer in the presence of organic (or other oxidizable) material than corresponding residuals of any of the other halogens.

Iodine reacts with water in the following manner:

$$I_2 + H_2O \leftrightharpoons HIO + H^+ + I^-$$
$$HIO \leftrightharpoons H^+ + IO^-$$

At pH values over 4.0, the hypoiodous acid (HIO) undergoes dissociation, as shown in the second equation. At pH values over 8.0, HIO is unstable and will not form the hypoiodite ion but will decompose according to the following equation:

$$3HIO + 2(OH^-) \leftrightharpoons HIO_3 + 2H_2O + 2I^-$$

This results in the formation of iodide (I$^-$) and iodate (HIO$_3$).

As the pH increases from 6.0 to 8.0, the percent of iodine as I$_2$ decreases while the percent of iodine as HIO increased. At pH 5.0 almost all the iodine is in the form of I$_2$. However, at pH 8.0, only 12 percent is present as I$_2$ while 88 percent is present as HIO and only 0.005 percent is present IO$^-$. Some concern has arisen over the possible harmful effects of iodinated water on thyroid function, on possible sensitivities to iodine itself, and objectionable tastes and odors.

However, field analyses of human users of water treated with iodine have failed to disclose any adverse effects on health. Results from the study of Black indicated that few subjects, if any, were able to detect by either color, taste or odor a concentration of 1.0 mg/l elemental iodine in water. Black also indicated that the iodide ion cannot be detected in concentrations exceeding 5.0 mg/l. When the concentration of iodine was between 1.5 and 2.0 mg/l, many people were able to detect a taste but found it unobjectionable.

Other workers showed iodine treated waters, at a dose of 8 mg/l added a faint to distinct taste to the water. A dose of 16 mg/l produced a taste and odor of distinct to decided. This was judged to be the upper limit of acceptability.

Since iodine does not combine with ammonia or many other organic compounds, the production of tastes and odors is minimized. Black states that iodine does not produce any detectable taste or odor in water containing several parts per billion phenol, whereas chlorine produces highly objectionable taste and odor under similar conditions.

Research into the bactericidal effects of iodine have shown only slight difference between pH 5.0 and pH 7.0. At pH values of 9 or above, a marked reduction in efficiency occurs. Apparently, hypoiodous acid formed at higher pH values is less bactericidal than free iodine. At pH 7.0, it has been found[48] that 99.99 percent kill of the most resistant of 6 bacterial strains tested could be achieved in 5 min with an free iodine residual of 1.0 mg/l.

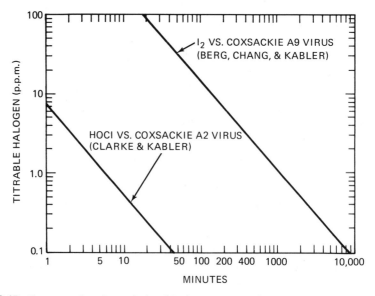

Fig. 6-15 Concentration time relationship for 99 percent destruction of two Coxsackie viruses by HOCl and I_2 at about 5°C.[28]

In general, it has been found that the bactericidal action of iodine resembled that of chlorine with respect to temperature and pH, but higher concentrations of iodine were required to produce comparable kills under similar conditions. However, toward higher organisms and especially toward cysts and spores, iodine and bromine may have some advantages.[49] Rapid (5–10 min) destruction of cysts with iodine concentrations of 5–10 mg/l have been reported for several different cysts in different neutral waters.

Figure 6-15 demonstrates the relative efficiencies of hypochlorous acid and elemental iodine in the destruction of 2 Coxsackie viruses. With equal weight concentrations of disinfectant, the iodine took 200 times longer to produce the same amount of destruction.

HIO (hypoiodous acid), which results from the hydrolysis of I_2 as the pH of solution increases above 6, destroys viruses at a rate considerably faster than that achieved by I_2. HIO_3 and I^- are essentially inert as viricides and I_3 has little or no viricidal activity.

In summary, iodine's great advantage over chlorine is that it does not react with ammonia or similar nitrogenous compounds. As a result, the bactericidal and cysticidal potency of I_2 and the viricidal effectiveness of HIO are not adversely affected by the presence of such nitrogenous pollutants, whereas the action of a given applied concentration of chlorine is badly affected either by exertion of demand or formation of halamines. Moreover, by appropriate control of dosage, pH, and other manageable conditions, aqueous iodine will occur about 50 percent as I_2 and 50 percent as HIO, thus providing a broad spectrum of germicidal capability.

However, iodine seems unlikely to become a municipal water supply disinfectant in any broad sense because of its high cost and restricted availability when compared to chlorine.

Bromine

As well-summarized by Morris,[33] bromine exhibits chemistry in water qualitatively similar to that of chlorine — hydrolyzing to HOBr, ionizing to OBr^-, and reacting with ammonia to form bromamines. The major differences are that the same degrees of ionization for HOBr occur at pH values about 1 pH unit greater than for HOCl and that bromamines appear to be nearly as active germicidally as HOBr. At the same time the bactericidal and viricidal effectiveness of HOBr has been found to be comparable to that of HOCl. As a result, solutions of free aqueous bromine exhibit a sharp decline in germicidal effectiveness with increasing pH only at pH greater than 8.5, whereas the effectiveness of free aqueous chlorine falls off sharply at pH greater than 7.5. In addition, the bactericidal and viricidal potency of bromine is reported to change in only a minor way in the presence of nitrogen in the form of ammonia, whereas aqueous bromine does undergo a rapid breakpoint process with ammonia, so that rapid bromine demand equal to about 20 times the ammonia-N may occur.

In spite of its advantages, it seems unlikely that bromine will become a general substitute for chlorine in municipal water disinfection. Its much greater cost and scarcity are strong drawbacks. Moreover, there have been no thorough investigations on a plant scale that would bring out engineering problems or indicate the general acceptability of waters treated with bromine. The fact that bromine or bromide is physiologically active to man in some measure may also be a bar to general acceptance.

Ozone

It has been reported[52] that "nearly a thousand cities purify their water with ozone," with the largest such plant being 238 mgd. Ozone is one of the two most potent and effective germicides used in water treatment. Only free residual chlorine can approximate it in germicidal power. The use of ozone for disinfection of municipal drinking water actually antedates chlorination.

The advantages of using ozone are its high germicidal effectiveness, which is the greatest of all known substances, even against resistant organisms such as viruses and cysts; its ability to ameliorate many problems of odor, taste, and color in water supplies; and the fact that upon decomposition, the only residual material is more dissolved oxygen. In addition, its potency is unaffected by pH or ammonia content.

Ozone must be produced at its point of usage by passing dry air between two high potential electrodes to convert oxygen into ozone. Improvements in the technology of ozone production have bettered the reliability and economy of its generation. The development of dielectric, ozone-resistant materials has simplified design and operation of ozone generators; in addition, new electronic designs have increased the efficiency of ozone production.

The earliest investigations of ozone for municipal water treatment took place in the last decade of the 19th century when it was shown ozone killed typhoid and cholera bacteria and the very resistant anthrax spore. Past research in the effectiveness of ozone as a germicide has been well summarized by Hann.[50] In general, ozone has been found to equal or exceed chlorine in its germicidal effects under a wide variety of circumstances.

Ozone has been found to be many times more effective than chlorine in inactivating the virus of poliomyelitis. Under experimental conditions, an identical dilution of the same strain and pool of virus, when exposed to chlorine in residual amounts of 0.5–1.0 ppm and to ozone in residual amounts of 0.05–0.45 ppm, was inactivated within 2 min by ozone, while 1 1/2–2 hr were required for inactivation by chlorine.

Ozone has also been reported to be several times faster in its germicidal effects on *Endamoeba histolytica*. For most cases, an ozone residual of 0.1 ppm for 5 min is adequate to disinfect a water low in organics and free of suspended material.[50] Organic material does exert a significant ozone demand.

The city of Paris, France now ozonates its water to such a degree that a minimum of 0.4 mg/l residual ozone is maintained for a 15 min contact. At this ozone concentration, hygienic security, including destruction of viruses, is equal to that provided by the best chlorination practice.

Nonetheless, the use of ozone does have disadvantages as summarized by Morris.[33] Because it must be produced electrically as it is needed and cannot be stored, it is difficult to adjust treatment to variations in load or to changes in raw water quality with regard to ozone demand. As a result, ozone historically has been found most useful for supplies with low or constant demand, such as groundwater sources. Moreover, although ozone is a highly potent oxidant, it is quite selective, and by no means universal in its action. Some otherwise easily oxidized substances, such as ethanol, do not react readily with ozone.

In some instances in which river waters rather heavily polluted with organic matter have been ozonated, results have indicated a rupture of large organic molecules into fragments more easily metabolized by microorganisms. This fragmentation, coupled with the inability of ozone to maintain an active concentration in the distribution system, led to increased slime growths and consequent deterioration of water quality during distribution.

In many ways, the desirable properties of ozone and chlorine as disinfectants are complementary. Ozone provides fast-acting germicidal and viricidal potency, commonly with beneficial results regarding taste, odor, and color. Chlorine provides sustained, flexible, controllable germicidal action that continues to be beneficial during distribution. Thus, it would seem that a combination of ozonation and chlorination might provide an almost ideal form of water supply disinfection. The procedure is being employed in a number of places, most particularly in the Netherlands at Amsterdam on Rhine River water. Ozone is used initially for phenol oxidation, viral destruction, and general improvement in physical quality; after normal water treatment, postchlorination is used to assure hygienic quality throughout the system. It seems logical that this approach will find increasing use in the U. S. as improvements in ozone generating equipment are made.

The costs of ozone, at the present time, are still higher than chlorine in the U. S. although there is potential for reductions in ozone generating costs. An analysis[53] of disinfection of secondary effluent from a 64 mgd plant was recently reported. The cost are summarized below:

Item	Liquid chlorine	Sodium hypochlorite	Ozone
Equipment, 15-yr[a] amort.	$ 6400	$ 330	$124,000
Building, 40-yr amort.	6600	7500	16,600
Chemical	82,000	186,000	40,200
Power	6800	700	88,000
Labor and maintenance	12,200	12,500	34,200
Total	$114,000	$210,000	$303,000
Cost, ¢/1000 gal	0.49	0.90	1.30

[a]13-yr for ozone generating equipment.

Although the costs above were developed for wastewater disinfection, similar cost differences are noted for water disinfection where ozonation typically costs about twice as much as simple chlorination[52] with ozonation operating costs typically being 0.1 to 0.2 cents/1000 gal for a dose of 1 mg/l, still a very small fraction of the overall costs of water treatment. For each pound of ozone produced and applied, 10 kw hr are consumed in large plants, with 12–15 kw hr in small ones.

Silver

A new approach[54] to applying the long-recognized disinfecting characteristics of silver is now commercially available. In this approach, silver ions are deposited on particles of granular activated carbon, a material which has an exceptionally high surface area for the deposit of silver. Bacteria-laden water contacting the silver impregnated carbon release minute quantities of silver (25 to 40 ppb) which serves as a disinfectant. The USPHS Drinking Water Standards set a mandatory limit of 50 ppb of silver in drinking water. Silver concentrations from these systems are reported to be within this limitation. The 50 ppb restriction is not based upon human toxicity; although silver is readily deposited in the body, it is not particularly toxic. The limitation is intended, however, to guard against argyrosis which is of cosmetic significance and which occurs only when high concentrations of silver are consumed.

Under normal or average operating conditions, the silver impregnated carbon is reported to be expected to treat 750,000 gal of raw water per ft^3. Equipment design is predicted on flow rate of 3 gal per min per ft^3 of material. A 10 to 30 min holding tank downstream of the column is recommended to assure enough contact time to complete disinfection. The advantages of this approach may be most pronounced in small installations where handling and control of chlorine compounds is a problem. Past attempts at using silver with other techniques have had limited use due to high cost, and the fact that silver may be bacteriostatic rather than bactericidal.

REFERENCES

1. Clarke, N. A., and Buelow, R. W., "Disinfection: What It Does," Proceedings, Fifth Annual Sanitary Engineering Symposium, California Department of Public Health, May, 1970.
2. "Water Chlorination — Principles and Practices, *American Water Works Association Manual of Water Supply Practices M-20*, 1973.
3. Melnick, J. L., "Detection of Virus Spread by the Water Route," Proceedings of Conference on Virus and Water Quality: Occurrence and Control, Dept. of Civil Engineering, University of Illinois, 1971.
4. Geldreich, E. E., and Clarke, N. A., "The Coliform Test: A Criterion for the Viral Safety of Water," Dept of Civil Engineering, University of Illinois, 1971.
5. McDermott, J. H., "Virus and Water Quality: Occurrence and Control — Conference Summary," Dept of Civil Engineering, University of Illinois, 1971.
6. Craun, G. F., and McCabe, L. J., "Review of the Causes of Waterborne-Disease Outbreaks," *Journal American Water Works Association*, p. 74 (January, 1973).
7. Berg, G., "Removal of Viruses from Water and Wastewater," Dept. of Civil Engineering, University of Illinois, 1971.
8. Oliver, D. O., "Viruses in Water and Wastewater: Effects of Some Treatment Methods," Dept. of Civil Engineering, University of Illinois, (1971).
9. Walton, G., "Effectiveness of Water Treatment Processes as Measured by Coliform Reduction," U. S. Public Health Service Publication No. 898 (1962).
10. Berg, G., Dean, R. B., and Dahling, D. R., "Removal of Poliovirus 1 from Secondary Effluent by Lime Flocculation and Sand Filtration," *Journal American Water Works Association*, p. 193 (February, 1968).
11. Sproul, O. J., "Virus Inactivation by Water Treatment," *Journal American Water Works Association*, p. 31 (January, 1972).
12. Thorup, R. T., Nixon, F. P., Wentworth, D. F., and Sproul, O. J., "Virus Removal by Coagulation with Polyelectrolytes," *Journal American Water Works Association*, p. 97 (February, 1970).
13. Chaudhuri, M., and Engelbrecht, R. S., "Removal of Viruses from Water by Chemical Coagulation and Flocculation," *Journal American Water Works Association*, p. 563 (September, 1970).
14. Manwaring, J. F., Chaudhuri, M., and Engelbrecht, R. S., "Removal of Viruses by Coagulation and Flocculation," *Journal American Water Works Association*, p. 298 (May, 1971).
15. Cookson, J. T., Jr., "Mechanism of Virus Adsorption on Activated Carbon," *Journal American Water Works Association*, p. 52 (January, 1969).
16. Robeck, G. G., Clarke, N. A., & Dostal, J. A. "Effectiveness of Water Treatment Processes in Virus Removal," *Journal American Water Works Association*, p. 1275 (October, 1962).
17. Hudson, H. E., "High Quality Water Production and Viral Disease," *Journal American Water Works Association*, p. 1265 (October, 1962).

18. Committee Report, "Viruses in Water," *Journal American Water Works Association*, p. 491 (October, 1969).

19. Laubusch, E. J., "Chlorine: Its Development, Characteristics and Utility for Disinfection and Oxidation," Proceedings, 3rd University of Illinois Sanitary Engineering Conference (1961).

20. American Water Works Association, *Water Treatment Plant Design*, New York, N. Y., (1969).

21. Buelow, R. W., and Walton, G., "Bacteriological Quality vs. Residual Chlorine, *"Journal American Water Works Association*, p. 29 (January, 1971).

22. Liu, O. C., Seraichekas, H. R., Akin, E. W., Brashear, D. A., Katz, E. L., and Hill, W. J., Jr., "Relative Resistance of Twenty Human Enteric Viruses to Free Chlorine in Potomac Water," Dept. of Civil Engineering, University of Illinois, (1971).

23. Burns, R. W., and Sproul, O. J., "Viricidal Effects of Chlorine in Wastewater," *Journal Water Pollution Control Federation*, p. 1834 (Nov., 1967).

24. Lothrop, T. L., and Sproul, O. J., "High Level Inactivation of Viruses in Wastewater by Chlorination," *Journal Water Pollution Control Federation*, p. 567.

25. Kruse, C. W., Olivieri, V. P., and Kawata, K., "The Enhancement of Viral Inactivation by Halogens," *Water and Sewage Works*, p. 187 (June, 1971).

26. Clarke, N. A., and Kabler, P. W., "The Inactivation of Purified Coxsackie Virus in Water by Chlorine," *American Journal of Hygiene*, p. 119 (January, 1954).

27. "Coliform Organisms as an Index of Water Safety, "Committee on Public Health Activities, *Journal of the Sanitary Engineering Division*, American Society of Civil Engineers, p. 41 (November, 1961).

28. Clarke, N. A., Berg, G., Kabler, P. W., and Chang, L. L., "Human Enteric Viruses in Water: Source, Survival, and Removeability," International Conference on Water Pollution Research, London (Sept., 1962).

29. Kabler, P. W., Chang, S. L., Clarke, N. A., and Clark, H. F., "Pathogenic Bacteria and Viruses in Water Supplies," 5th Sanitary Engineering Conference, U. of Illinois, Jan., 1963.

30. Berg, G., "Virus Transmission by the Water Vehicle III. Removal of Viruses by Water Treatment Porcedures," *Health Laboratory Science*, p. 17 (July, 1966).

31. Kruse, C. W., Olivieri, V. P., and Kawata, K., "The Enhancement of Viral Inactivation of Halogens," Dept of Civil Engineering, University of Illinois, 1971.

32. Collins, H. F., "Chlorination — How to Use It," Proceedings Fifth Annual Sanitary Engineering Symposium, California Department of Public Health (May, 1970).

33. Morris, J., "Chlorination and Disinfection-State of the Art," *Journal American Water Works Association*, p. 769 (December, 1971).

34. Michalek, S. A., and Leitz, F. B., "On-Site Generation of Hypochlorite," *Journal Water Pollution Control Federation*, p. 1697 (Sept., 1972).

35. "Cheaper Sewage Treatment," *Chemical Week*, p. 52 (January 12, 1972).

36. Baur, F., "Contract Granted for Sodium Hypochlorite Generator," *Public Works*, p. 76 (June, 1972).

37. "Chloropac," Engelhard Industries brochure.

38. "Sanilec Systems," Diamond Shamrock brochure.

39. "Cloromat," Ionics, Inc. brochure.

40. "Pep-Clor Systems," Pacific Engineering and Production Co. of Nevada brochure.

41. Laubusch, E. J., "Hypochlorites," *Public Works* p. 106, (June, 1963).

42. Stenquist, R. J., and Kaufman, W. J., "Initial Mixing in Coagulation Processes," EPA Report EPA-R2-72-053 (November, 1972).

43. Scarpino, P. V., Berg, G., Chang, S. L., Dahling, D., and Lucas, M., "A Comparative Study of the Inactivation of Viruses in Water by Chlorine," *Water Research*, p. 959 (1972).

44. Stringer, R., and Kruse, C. W., "Amoebic Cysticidal Properties of Halogens," *Proceedings of the National Specialty Conference on Disinfection*, American Society of Civil Engineers, New York, N. Y. (1970).

45. Connell, G. F., and Fetch, J. J., "Advances in Handling Gas Chlorine," *Journal Water Pollution Control Federation*, p. 1505 (August, 1969).

46. Black, A. P., Kinman, R. N., Thomas, W. C., Jr., Freund, G., and Bird, E. D., "Use of Iodine for Disinfection," *Journal American Water Works Association*, p. 140 (Nov., 1965).

47. Black, A. P., Thomas, W. C., Jr., Kinman, R. N., Bonner, W. P., Keirn, M. A., Smith J. J., Jr., and Jabero, A. A., "Iodine for the Disinfection of Water," *Journal American Water Works Association*, p. 69 (Jan., 1968).

48. Karalekas, P. C., Jr., Kuzminski, L. N., and Feng, T. H., "Recent Developments in the Use of Iodine for Water Disinfection," *Journal New England Water Works Association*, p. 152 (June, 1970).

49. McKee, J. E., Brokaw, C. J., and McLaughlin, R. T., "Chemical and Colicidal Effects of Halogens in Sewage," *Journal Water Pollution Control Federation*, p. 795 (Aug., 1960).

50. Hann, V. A., "Disinfection of Drinking Water With Ozone," *Journal American Water Works Association*, p. 1316 (October, 1956).

51. "Parisian Water Gets Ozone Treatment," *Water and Wastes Engineering*, p. 83 (Sept., 1969).

52. Bendes, R. J., "Ozonation, Next Step to Water Purification," *Power*, (August, 1970).

53. Yao, K. M., "Is Chlorine the Only Answer," *Water and Wastes Engineering*, p. 30 (January, 1972).

54. "Hygiene Water Purifiers," brochure, Ionics, Inc.

7 Heavy Metal Removal

INTRODUCTION

As discussed in Chapter 1, heavy metal pollution has recently become a topic of general public concern as a result of widespread publicity on mercury pollution as well as on other heavy metals. The following list of heavy metals has been classified[1] as having very high or high pollution potential:

1. Very high pollution potential: Ag, Au, Cd, Cr, Cu, Hg, Pb, Sb, Sn, Te, Zn.
2. High pollution potential: Ba, Bi, Fe, Mn, Mo, Ti, U.

The pollution potential of these metals is based on toxicity to individual species in a broad spectrum of living things; however, the level of toxicity is not clearly defined in many cases. Toxicity of heavy metals to human beings is reflected in the drinking water standards issued by the U. S. Public Health Service. A list of the standards for the several of the above heavy metals is given in Table 7-1. It must be noted, however, that the standards are not based on toxicity alone, for some metals, notably Fe and Mn, which are objectionable with respect to taste of the water and the color imparted to laundered goods washed in the water.

This chapter is concerned with the reduction in concentration of these metals by a variety of commonly used unit processes such as

TABLE 7-1 HEAVY METAL CONCENTRATIONS — U. S. PUBLIC HEALTH SERVICE DRINKING WATER STANDARDS (1962 REVISION)

Substance		Maximum Concentration in mg/l
Arsenic	(As)	0.05
Barium	(Ba)	1.0
Cadmium	(Cd)	0.01
Chromium	(Cr^{+6})	0.05
Copper	(Cu)	1.0
Iron	(Fe)	0.3
Lead	(Pb)	0.05
Manganese	(Mn)	0.05
Selenium	(Se)	0.01
Silver	(Ag)	0.05
Zinc	(Zu)	5.0

coagulation and settling, sand or mixed-media filtration, and carbon adsorption. The precipitation of metal hydroxides is governed by the concentration of the metal ion in solution and the pH. In general, as the pH increases, the solubility of the metal hydroxide decreases. This relationship is expressed in the equation for the solubility product of a compound:

$$(M^{+z})\,(OH)^2 = K_{sp} \text{ (solubility product)}$$

The solubility products of cationic heavy metal oxides or hydroxides are listed in Table 7-2.

The logarithm for the sum of the concentration of free metal ions and the concentration of hydroxide complexes has been calculated[18] as a function of pH from values of the solubility products and complex constants of the hydroxides taken from literature is shown in Figure 7-1. The lines showing those areas where the content is greater than 1 mg/l are marked with heavier lines. The slope of the lines increases by one unit where the first hydroxide complex begins to dominate the contents of the free metal ions and with a further unit where the second complex begins to dominate the first.

If the rise in pH is brought about with NaOH, then the solubility of Cr (III) will rise again, when the pH value rises above that of approximately 8 and of Zn (II) when it rises above approximately 9. If the rise in pH is brought about by lime, then a rise in solubility does not occur as the solubilities of calcium zincate and calcium chromite are relatively low.

The curves show that Pb(II), Cu(II), and Cr(III) are precipitated to the concentration of 1 mg/l at a pH value of about 7. These correspond to

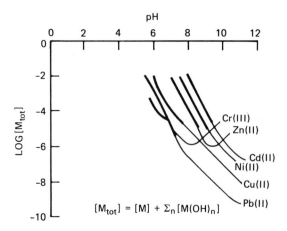

Fig. 7-1 The solubility of pure metal hydroxides as a function of pH.[18] Heavy portions of lines show where concentrations are greater than 1 mg/l.

the values obtained by precipitation with aluminium sulfate. Experimental results[18] concerning Zn(II), Cd(II), and Ni(II) were, however, much better than those shown in the diagram at pH 7 because the metals were coprecipitated with aluminium hydroxide.

TABLE 7-2 SOLUBILITY PRODUCTS OF CATIONIC HEAVY METAL OXIDES OR HYDROXIDES[4,5,6]

Compound	K_{sp}
Ag_2O	2×10^{-8}
$Au(OH)_3$	8.5×10^{-45}
$BaCO_3$[a]	1.6×10^{-9}
$BaSO_4$[a]	1×10^{-10}
$BiOOH$	3×10^{-11}[b]
$Cd(OH)_2$	2×10^{-14}
$Cr(OH)_3$	1×10^{-30}
$Cu(OH)_2$	3×10^{-19}
$Fe(OH)_2$	1.8×10^{-15}
$Fe(OH)_3$	6×10^{-38}
HgO	3×10^{-26}
$Mn(OH)_2$	2×10^{-13}
$Pb_2O(OH)_2$	1.6×10^{-15}
$Sn\ O$	1×10^{-61}
$Ti\ (OH)_3$	1×10^{-40}[b]
Zn	4.5×10^{-17}

[a]Barium compounds that will form preferentially to the hydroxide in most waters.
[b]Estimated value, not verified.

LIME COAGULATION

Since many of the trace metals form insoluble hydroxides as oxides at pH 11, lime coagulation results in a reduction of these metal concentrations. It appears that it is possible to reduce some of the metal concentrations below that predicted by the solubility products. This may be due to adsorption of the metal ions by the chemical floc. On the other hand, when any of the metals exist in organic form, the concentration reduction during lime coagulation may be less than expected from solubility considerations. Table 7-3 summarizes the effects of lime coagulation on a number of heavy metals. Some of these data were collected on industrial metal wastes which have metal ion concentrations a great deal higher than occur in any municipal water treatment plant influent. These data were included due to the scarcity of metal reduction results from actual chemical coagulation of water supplies.

After lime coagulation, filtration to remove residual particulate matter will provide additional removal of heavy metals. Table 7-4 gives the results of sand filtration of the same wastes considered in the lime coagulation section of this chapter.

CARBON ADSORPTION

With water containing no organics, carbon contacting usually has little effect on metal ions. It has been found,[10] however, for several metal ions that activated carbon treatment of wastewaters results in a surprising reduction (over 96 percent for 3 of the 4 metal ions studied). The cause of this is not clearly identified, but it appears possible that organics may serve as coadsorbates linking the metal ions and the carbon. The above study did not include thermal regeneration of the carbon, so it is not known at this time what effect regeneration would have upon the capacity for adsorption of metal ions. Activated carbon has been used to remove un-ionized species such as arsenic and antimony from an acidic stream.

ALUM COAGULATION

Alum coagulation, as noted earlier, will provide removal of some heavy metals at near neutral pH values. Table 7-6 summarizes the removals noted by one observer.[18]

Alumina Adsorption

Arsenic may be removed by passing the water through a bed of 28 × 48 mesh granular activated alumina.[17] The removals provided are much better than that which can be achieved by alum coagulation.

TABLE 7-3   REMOVAL OF HEAVY METALS BY LIME COAGULATION AND RECARBONATION

Metal	Reference	Concentration Before Treatment mg/l	Concentration After Treatment mg/l	Final pH	% Removal
Antimony[a]	5	—	—	11	90
Arsenic[a]	5	—	—	11	<10
Barium[a]	18	23	23	9.5	0
Bismuth[a]	5		~1.3 (sol)[b]	11	
	5		.0002 (sol)	11	
Cadmium	9	Trace		11	~50
Chromium (+6)	10	0.0137	0.00075	>11	94.5
Chromium (+3)	10	0.056	0.050	>11	11
	11	7,400	2.7	8.7	99.9+
	18	15	0.4	9.5	97
Copper	11	15,700	0.79	8.7	99.9+
	12	7	1	8	86
	12	7	.05	9.5	93
	13	302	Trace	9.1	99+
	18	15	0.6	9.5	97
Gold[a]	5		<.001 (sol)	11	90+
Iron	13	13	2.4	9.1	82
	14	17	0.1	10.8	99+
	14	2.0	1.2[c]	10.5	40
Lead[a]	5	—	<.001 (sol)[b]	11	90+
	18	15	0.5	9.5	97
Manganese	14	2.3	<0.1	10.8	96
	14	2.0	1.1[c]	10.5	45
	15	21.0	0.05	9.5	95
Mercury[a]	5		Oxide soluble		<10
	9	Trace	—	8.2	~10
Molybdenum	18	11	9	9.5	18

Nickel	11	160	0.08	8.7	99.9+
	12	5	0.5	8.0	90
	12	5	0.5	9.5	90
	16	100	1.5	10.0	99
	18	16	1.4	9.5	91
Selenium	10	0.0123	0.0103	>11	16.2
Silver	10	0.0546	0.0164	>11	97
Tellurium[a,d]	5		(<0.001?)	11	(?90+)
Titanium[a,d]	5		(<0.001?)	11	(?90+)
Uranium[e]	5		?		?
Zinc	5	17	.007 (sol)	11	90+
	18		0.3	9.5	98

[a]The potential removal of these metals were estimated from solubility data.
[b]Barium and lead reductions and solubilities are based upon the carbonate.
[c]These data were from experiments using iron and manganese in the organic form.
[d]Titanium and tellurium solubility and stability data made the potential reduction estimates unsure.
[e]Uranium forms complexes with carbonate ion. Quantitative data were unavailable to allow determination of this effect.

TABLE 7-4 HEAVY METAL REMOVAL BY FILTRATION FOLLOWING LIME COAGULATION

Metal	Reference	Concentration Before Treat.	Concentration After Treat.	pH	% Removal
Cd	9	Trace		8.1	95.0
	10	0.00075 mg/l	0.00070	7.6	6.6
Cr^{+6}	10	0.0503	0.049	7.6	2.6
Cr^{+3}	11	2.7	0.63	8.7	77.0
Cu	11	0.79	0.32		59.5
	12	—	.5	9.5	
Fe	14		0.1	10.8	
	14		1.2 Organic	10.5	
Mn	14		<0.1	10.8	
	14		1.1 Organic	10.5	
Ni	11	0.08	0.1	8.7	
	12		0.5	9.5	
Se	10	.0103	0.00932	>11	9.5
Ag	10	0.00164	0.00145	>11	11.6
Zn	11	0.97	0.23	8.7	76.3
	12		2.5	9.5	

Tests on a water containing an arsenic concentration above 0.1 mg/l and fluoride of 0.7 mg/l (fluoride is also removed by the alumina) indicated that 7000 gallons of water could be treated per ft^3 of alumina while maintaining the arsenic concentration at values less than 0.01 mg/l. Surface loading rates on the alumina bed of 2.5–3 gpm/ft^2 were used. Regeneration was achieved by using 4 bed volumes of a 1 percent solution of sodium hydroxide followed by a rinse with 8 bed volumes of raw water. At least 1 bed volume of 0.05 N sulfuric acid was then used as an acid rinse followed by a final rinse with at least 1 bed volume of raw water. The cost of such treatment was estimated at $15–$50/mg, depending upon the amount of fluoride present. As the fluoride concentration decreases, the costs are reduced toward the lower end of the above range. The process can use readily available standard ion-exchange equipment which can be automated to reduce operator attention to a minimum.

TABLE 7-5 HEAVY METAL REMOVAL BY CARBON ADSORPTION

Metal	Reference	pH Before Carbon	Concentration Before Carbon	Concentration After Carbon	% Removal
Cd	10	7.6	0.00070	9×10^{-6}	98.7
Cr^{+6}	10	7.6	0.049	0.00171	96.5
Se	10	7.6	0.00932	0.00585	37.2
Ag	10	7.6	0.00145	0.000048	96.7

TABLE 7-6   REMOVAL OF HEAVY METALS BY ALUM COAGULATION AT A
pH OF 6.8–7.0[18]

Metal	Concentration Before Treatment, mg/l	Concentration After Treatment, mg/l
Lead	17	1.3
Copper	15	1.7
Molybdenum	11	9.0
Chromium (+ 6)	3	2.3
Chromium (+ 3)	15	0.2
Zinc	17	11.0
Nickel	16	17.0

COMBINED PROCESSES

A report[19] of the efficiency of heavy metal removal from
sewage (secondary effluent) at the Orange County Water District
(California) was recently published. The Orange County Water
District's wastewater reclamation pilot plant consisted of units for
chemical treatment, sedimentation, ammonia stripping, recarbonation,
mixed-media filtration, activated carbon adsorption and chlorination.
During the operation of the pilot plant, monthly composite samples of
the plant's influent and effluent were collected. Each month, these
samples were sent to the California State Health Department where the
analyses for several heavy metals were performed. Typical water quality data indicating the performance and efficiency of the pilot wastewater reclamation plant are given in Table 7-7.

During the 9 month test period, the arsenic concentration in the
plant's influent varied between 0.00–0.01 mg/l. The plant's effluent
concentration varied from 0.00–0.03 mg/l. The pilot plant treatment
system was not particularly consistent in removing arsenic; however,
this is probably a result of the extremely low concentration found in the
plant's influent. During 2 of the 9 months, the plant did remove 100
percent of the arsenic.

The results indicate that there is little danger of a public health
hazard from arsenic in a properly designed wastewater reclamation
plant's effluent. Concentrations of arsenic found in the plant's influent
and effluent were always below the regulatory agencies' requirements
of 0.05 mg/l.

The pilot plant influent usually contained no barium. Only once
during the test period was any barium detected, and its concentration
was less than 0.02 mg/l.

Cadmium concentrations found in the plant's influent ranged from a
low of 0.011 mg/l to a high of 0.022 mg/l. Following treatment, effluent
concentrations were reduced to a range of 0.00 mg/l to 0.005 mg/l. The

TABLE 7-7   TYPICAL WATER QUALITY PILOT WASTEWATER RECLAMATION PLANT ORANGE COUNTY, CALIFORNIA

Constituent	Concentration, mg/l	
	Influent	Effluent
Calcium	70–110	80
Magnesium	20–45	2
Sodium	240–260	240–260
Potassium	20–35	20–35
Bicarbonate	200–450	250
Sulfate	270–350	270–350
Chloride	300–350	300–350
Phosphate	20–25	1
Nitrogen		
Organic	5–15	1
Ammonia	15–30	2
Nitrite	1	1
Total dissolved solids	1200–1400	1000–1100
Suspended solids	30–80	1
BOD	30–80	2
COD	100–200	10–30
MBAS	3–4	0.1

operating efficiency of the plant for removing cadmium varied between 74–100 percent. The average removal was 89 percent. Higher efficiency appears to be a function of influent concentration. As influent concentration increases, so does removal. During the test period, the cadmium concentration in the pilot plant's influent was never below 0.01 mg/l limit set by the regulatory agencies; but after treatment, the pilot plant's effluent was always well below the 0.01 mg/l limit.

The concentration of chromium in the plant influent ranged from 0.09 to 0.19 mg/l. After treatment, the pilot plant effluent contained only 0.01 to 0.04 mg/l of chromium. The pilot treatment processes were very effective in removing chromium. Removal efficiencies varied from 89 to 100 percent. During the test period, the pilot plant influent always contained concentrations of hexavalent chromium that exceeded the 0.05 mg/l limit. After treatment, however, the concentration was always reduced below the permissible limit.

The concentration of lead in the plant's influent varied from 0.00 to 0.05 mg/l. Following treatment, these concentrations were reduced to a range of 0.00 to 0.04 mg/l. The removal efficiency varied from 0 to 50 percent. The average removal was 30 percent. The treatment processes were not particularly effective or consistent in removing lead from the wastewater, probably due to the very low concentrations of lead in the influent. During the test period, both the pilot plant influent and effluent always contained concentrations of lead within the imposed limit of 0.05 mg/l.

The concentration of selenium in the plant's influent varied from 0.00 to 0.003 mg/l. The concentration of selenium in the plant's effluent was 0.00 to 0.003 mg/l. A large variation in removal efficiency was observed. Removals ranged from 0–89 percent. The highest efficiency was observed when the influent concentration of selenium was a maximum. It appears that higher efficiencies are increasingly difficult to achieve at lower concentrations. During the test period, neither the plant's influent nor effluent contained concentrations of selenium that exceeded the imposed limit of 0.01 mg/l.

The pilot treatment processes were not very consistent in removing silver, even though during 1 month 100 percent of the silver was removed. This inconsistency was probably a result of the extremely low influent concentrations. During the test period, silver was found in the plant influent only 3 times. Neither the plant's influent nor effluent ever exceeded a silver concentration of 0.01 mg/l and was always below the maximum allowable silver concentration of 0.05 mg/l imposed by the regulatory agencies.

The mercury concentration in the plant influent was between 0.001 and 0.003 mg/l. The plant effluent concentration varied from 0.00 to 0.006 mg/l. The pilot plant was inconsistent in further reducing the extremely low influent mercury concentration. The maximum concentration of mercury permitted by the discharge requirements was 0.005 mg/l. During the test period, the pilot plant influent was always below this allowable concentration.

The copper concentration in the plant's influent varied from 0.09 to 0.39 mg/l. The concentration of copper in the plant's effluent varied from 0.02 to 0.30 mg/l. The removal efficiency ranged from 56 to 95 percent. The average removal was 70 percent. The highest removal occurred when influent copper concentration was maximum. During the test period, neither the influent nor effluent copper concentration exceeded the limit of 1.0 mg/l.

The zinc concentration in the plant influent varied from 0.07 to 2.08 mg/l and the zinc concentration in the effluent was between 0.02 and 0.07 mg/l. The removal efficiency varied from 56 to 99 percent. The highest removal occurred when influent concentration was maximum. During the test period, neither the influent nor effluent zinc concentration exceeded the imposed limit of 5.0 mg/l.

The results of this investigation indicate that certain heavy metals can be removed along with organics, suspended solids, and other inorganics by lime coagulation, mixed-media filtration, and activated carbon adsorption. The results of actual full-scale advanced wastewater treatment plant operation in removing heavy metals are given below in Table 7-8, for the Lake Tahoe advanced wastewater treatment plant. The treatment processes employed as described earlier are similar to those used in Orange County. The principal difference is that the Tahoe

TABLE 7-8   IONIC ANALYSES FOR METALS IN RECLAIMED WATER

South Tahoe Public Utility District, 1971
(Tests by Battelle—NW Research)

Element	Reclaimed Water	Maximum Allowable U.S.P.H.S. Drinking Water Standards Concentration in mg/l
Arsenic	0.005	0.05
Chromium^{+6}	0.0005	0.05
Copper	0.0116	1.0
Iron	0.0003	0.3
Manganese	0.002	0.05
Selenium	0.0005	0.01
Silver	0.0004	0.05
Zinc	0.005	5.0
Bromine	0.065	
Uranium	0.0015	
Cobalt	0.00033	
Cesium	0.000006	
Mercury	0.0005	
Rubidium	0.010	
Scandium	0.000001	
Antimony	0.00044	

wastewater is entirely of domestic origin, while the Orange County wastes included some from industrial operations.

Note that heavy metal concentrations are all well below the maximum limits specified in the Drinking Water Standards after lime coagulation, settling, filtration, and granular carbon adsorption.

REFERENCES

1. Pringle, B. H., Hissange, D. E., Katz, E. L., and Mulawka, S. T., "Trace Metal Accumulation by Estuarine Mollucks," *J. San. Eng. Div.*, **94**, p. 455–473 (1968).
2. Jenkins, S. H., Keight, D. G., and Ewins, A., "The Solubility of Heavy Metal Hydroxides in Water, Sewage, and Sewage Sludge — II, The Precipitation of Metals by Sewage," *Int. J. Air. Wat. Poll.*, **8**, p. 679–693, (1964).
3. Jenkins, S. H., Keight, D. G., and Humphrey, R. E., "The Solubility of Heavy Metal Hydroxides in Water, Sewage, and Sewage Sludge — I, The Solubility of Some Metal Hydroxides," *Int. J. Air Wat. Poll.*, **8**, p. 537–556 (1964).
4. Homer, Jackson, Thurston, *Industrial Waste Treatment Practice*, Butterworths associated with Imp. Chem. Ind. Ltd., (1961).
5. Latimer, W. M., *The Oxidation States of the Elements and Their Potentials*

in Aqueous Solutions, 2d ed. Prentice-Hall, Inc. Englewood Cliffs, N. J., 1952.

6. Feitknecht, W., and Schindler, P. "Solubility Constants of Metal Hydroxide Salts in Aqueous Solution," IUPAC *Pure and Applied Chemistry,* **6,** p. 180 (1963).

7. Barth, E. F., Ettiger, M. B., Salotto, B. V. and McDermott, J. N., "Summary Report on the Effects of Heavy Metals on Biological Treatment Processes," *Journal Water Pollution Control Federation,* **37,** p. 1, 86 (Jan., 1965).

8. Barth, E. F., English, J. N., Salotto, B. V., Jackson, B. N., and Ettinger, M. B., "Field Survey of Four Municipal Wastewater Treatment Plants Receiving Metallic Wastes," *Journal Water Pollution Control Federation,* **37,** p. 1101, 1121 (Aug. 1965).

9. *Report of the Joint Program of Studies on the Decontamination of Radioactive Waters,* ORNL-2557, TID-4500 (14th ed.) Oak Ridge National Laboratory, Robert A. Taft Sanitary Engineering Center, Feb., 1959.

10. Linstedt, K. D. and J. T. O'Connor, University of Colorado and Illinois, *Behavior of Trace Elements in Water Reclamation Treatment Processes,* Build Cooperative Research Report (Aug., 1969).

11. Stone, E. H. F., "Treatment of Non Ferrous Metal Process Wastes of Kynoch Works, Birmingham, England," *Proceedings, 22 Annual Industrial Waste Conference,* Purdue University, Part 2, p. 848 (1967).

12. Walker, C. A., *et al.,* "Treatment of Lagoon Waters," *Sewage and Industrial Wastes,* Vol. **26,** No. 8, p. 1008 (Aug., 1954).

13. McElhoney, H. W., "Metal-Finishing Wastes Treatment at the Meadville, Pa. Plant of Talon, Inc.," *Sewage and Industrial Wastes,* Vol. **25,** No. 4, p. 475 (April, 1953).

14. Nordell, Eskel, *Water Treatment for Industrial and Other Uses,* 2d ed. Reinhold Pub. Corporation, New York, 1961.

15. Robinson, Jr., L. R., and R. I., Dixon, "Iron and Manganese Precipitation in Low Alkalinity Ground Waters," *Water and Sewage Works* (Nov. 1968).

16. Kantawala, D., and H. D. Tomlinson, "Comparative Study of Recovery of Zn and Ni by Ion Exchange Media and Chemical Precipitation," *Water and Sewage Works* R280 (1964).

17. Bellack, E., "Arsenic Removal from Potable Water," *Journal American Water Works Association,* p. 454 (July, 1971).

18. Nilsson, R., "Removal of Metals By Chemical Treatment of Municipal Wastewater," *Water Research,* **5,** p. 51, (1971).

19. Wesner, G. M., Argo, D., and Culp, G. L., "Heavy Metals Removal in Wastewater Treatment Processes," Orange County Water District Report (1972).

20. Argo, D. G., and Culp, G. L., "Heavy Metals Removal in Wastewater Treatment Processes," *Water and Sewage Works,* p. 62 (August, 1972) and p. 128 (September, 1972).

8 Activated Carbon Treatment

PRINCIPLES OF CARBON ADSORPTION

The use of powdered carbon applied ahead of water treatment plant clarifiers or filters for taste and odor control is a long established procedure and will not be discussed further in this book (see Chapter III for discussion relative to feed of powdered carbon directly to mixed-media filters). Instead, the emphasis will be on recent developments in the integration of the adsorption and filtration processes through the incorporation of granular carbon in the filter bed itself. The well-established powdered carbon techniques will not be completely displaced by those described in this chapter as the powdered carbon approach may offer economic advantages when very low or infrequent carbon usage is required to solve a specific problem.[10]

Activated carbon removes organic contaminants from water by the process of adsorption which results from the attraction and accumulation of one substance on the surface of another. In general, the chemical nature of the carbon surface is of relatively minor significance in the adsorption of organics from water and secondary to the magnitude of the surface area of carbon available. For this reason, high surface area is the prime consideration in adsorption. Granular activated carbons typically have surface areas of 500–1400 m²/g. Activated carbon has a preference for organic compounds and, because of this selectivity, is particularly effective in removing organic compounds which may cause taste and odor problems in water supplies. In addition to the taste and odor problems trace concentrations of organics in drinking water supplies may have adverse health effects over the long term, as discussed in more detail elsewhere in this book.

Activated carbon can be made from a variety of materials such as coal, wood, nut shells, and pulping waste. Activated carbon made from wood is normally termed charcoal and was the first type of granular carbon to find its way into municipal water treatment.

Granular carbons made from coal are hard and dense, and can be pumped in a water slurry without appreciable deterioration. Hydraulic handling of carbon allows dust-free loading and unloading of filters. Coal-derived granular carbons are well-suited to water treatment because the carbon wets rapidly and does not float, but conforms to a neatly packed bed with acceptable pressure drop characteristics. Also, the carbon is quite dense, and thereby makes available more adsorption capacity in any given filter volume. Much of the surface area available for adsorption in granular carbon particle is found in the pores within the granular carbon particles created during the activation process.

The major contribution to surface area is located in pores of molecular dimensions. A molecule will not readily penetrate into a pore smaller than a certain critical diameter and will be excluded from pores smaller than this; thus, molecules are "screened out" by pores smaller than a minimum diameter which is a characteristic of the adsorbate and related to molecular size. Furthermore, for any molecule the effective surface area for adsorption can exist only in pores which the molecule can enter.

Figure 8-1 illustrates this concept for the case in which two adsorbate molecules in a solvent (not shown) compete with each other for adsorbent surface. Because of the irregular shape of both pores and molecules and also by virtue of constant molecular motion, the fine pores are not blocked by the large molecules but are still free for entry by small molecules. As a contributing factor, the greater mobility of the smaller molecule should permit it to diffuse ahead of the larger molecule and penetrate the fine pores first. The forces of attraction between the carbon and adsorbing molecules are known to be greater the more similar the adsorbing molecules are in size to the pores. The most tenacious adsorption takes place when the pores are barely large enough to admit the adsorbing molecules. The smaller the pores with respect to the molecules, the greater the forces of attraction. The pores cannot be so small, however, that the adsorbing molecules find it difficult to enter, or the adsorptive for those molecules will be greatly reduced. The pore structure of activated carbons is extremely important in determining their adsorptive properties.

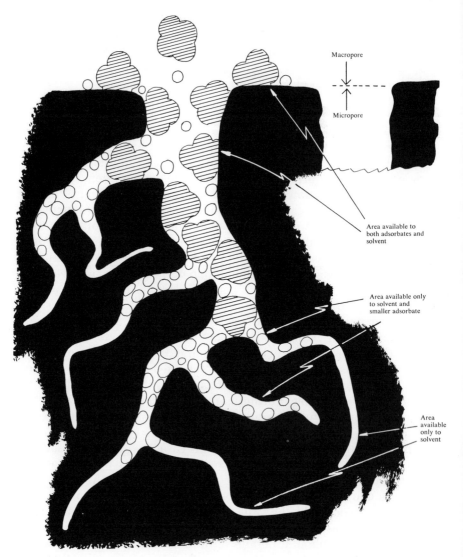

Macropore

Micropore

Area available to both adsorbates and solvent

Area available only to solvent and smaller adsorbate

Area available only to solvent

Fig. 8-1 Concept of molecular screening in micropores. *(Courtesy Calgon Corp.)*

MANUFACTURE OF GRANULAR CARBONS

Activated carbons can be made from a variety of carbonaceous raw materials. Generally, the process consists of selection of an organic raw material, dehydration and carbonization, and finally, activation to produce a highly porous structure. Raw materials that have been used to produce activated carbons include wood, lignin, nutshells, coal, lignite, peat, bagasse, sawdust, and petroleum residues. Granular carbons prepared from coal generally have the best physical characteristics (specifically, hardness and density) for use in water treatment filters. The better grades of powdered carbons are made from lignin and lignite. The dehydration and carbonization step is usually accomplished by the slow heating of the raw material in the absence of air. Sometimes a dehydrating agent such as zinc chloride or phosphoric acid is used. The activation step may be carried out using chemicals in the solid or liquid state, or by using mixtures of gases. The more modern techniques use mixtures of oxidizing gases such as steam, air, and CO_2.

Activated carbons are usually classified according to their physical state, e. g., powdered or granular, and according to their use, e.g., water grade, decolorizing, liquid phase, or gas phase. Granular carbons are those materials which are greater in particle size than approximately 150 mesh. Powdered carbons are those which are smaller in particle size.

DESIGN CONSIDERATIONS

Characterization of Granular Carbons

The important characteristics of a carbon must be expressed both in terms of adsorptive characteristics and physical properties. The adsorptive capacity of a carbon can be measured to a fair degree by determining the adsorption isotherm experimentally in the system under consideration. Simpler capacity tests such as the Iodine Number or the Molasses Number also may be used as a measure of adsorptive capacity.

The adsorption isotherm is the relationship, at a given temperature, between the amount of a substance adsorbed and its concentration in the surrounding solution. If a color adsorption isotherm is taken as an example, the adsorption isotherm would consist of a curve plotted with residual color in the water as the abscissa, and the color adsorbed per gram of carbon as the ordinate. A reading taken at any point on the isotherm gives the amount of color adsorbed per unit weight of carbon, which is the carbon adsorptive capacity at a particular color concentration and water temperature. In very dilute solutions, such as water supplies, a logarithmic isotherm plotting usually gives a straight line.

In this connection, a useful formula is the Freundlich equation, which relates the amount of impurity in the solution to that adsorbed as follows:

$$\frac{x}{m} = kC^{1/n}$$

where x = amount of color adsorbed

m = weight of carbon

$\frac{x}{m}$ = amount of color adsorbed per unit weight of carbon

k and n are constants

C = unadsorbed concentration of color left in solution

In logarithmic form:

$$\log \frac{x}{m} = \log k + \frac{1}{n}\log C$$

in which $1/n$ represents the slope of the straight line isotherm. Detailed procedures for establishing the experimental conditions and conducting and interpreting isotherm adsorption tests are presented in detail elsewhere.[7,8]

From an isotherm test, it can be determined whether or not a particular purification can be effected. It will also show the approximate capacity of the carbon for the application, and provide a rough estimate of the carbon dosage required. Isotherm tests also afford a convenient means of studying the effects of pH and temperature on adsorption. Isotherms put a large amount of data into concise form for ready evaluation and interpretation. Isotherms obtained under identical conditions using the same test solutions for two test carbons can be quickly and conveniently compared to reveal the relative merits of the carbons.

For some applications, it is not necessary to know the complete adsorption isotherm to determine the best carbon for a particular application or to specify the most appropriate carbon. In these cases, it is possible to use simpler capacity tests such as the Iodine Number or Molasses Decolorizing Index as an appropriate measure of adsorptive capacity.

The Iodine Number is the number of milligrams of iodine adsorbed per gram of carbon when in equilibrium, under specified conditions, with a solution of 0.02 N iodine concentration.[7,8]

The Iodine Number is an approximate measure of the adsorptive capacity of a carbon for small molecules such as iodine. The Molasses Decolorizing Index is, roughly, a measure of the adsorptive capacity of the carbon for color bodies in a specified molasses solution, as compared to a standard carbon. The Molasses Decolorizing Index is a measure, therefore, of the adsorptive capacity for large molecules, such as color bodies.

The physical properties of granular carbons which are important in performance include resistance to breakage, particle size, and degree of dustiness. For measuring resistance to breakage there are empirical tests such as Abrasion Number and Hardness Number. Particle size is determined by a screen analysis from which mean particle diameter and effective size can be calculated. Density is simply the weight per unit volume of the carbon. Typical specifications of a granular carbon suitable for water treatment filter applications are as follows:

Total surface area (m²/g)	850–1500
Bulk density (lb/ft³)	26
Particle density, wetted in water (g/cc)	1.3–1.4
Effective size (mm)	0.8–0.9
Uniformity coefficient	1.9 or less
Mean particle diameter (mm)	1.5–1.7
Iodine number	Min 850
Abrasion number	Min 70
Ash	Max 8%
Moisture	Max 2%

For incorporation in filter designs, carbon particle size distribution is very important. If the carbon is to replace anthracite coal in a dual media filter, it should have similar filtration characteristics and similar backwashing characteristics. Fortunately, as shown below, commercial carbons are available with characteristics which are very similar to anthracite coals used as filter media.

TABLE 8-1 COMPARISON OF COAL AND GRANULAR CARBON AS FILTRATION MEDIA

	Values for Hard Coal Media	Filtrasorb[a] 8 × 30 Mesh[b]	14 × 40 Mesh[b]
Real density g/cc	1.5–1.6	2.1	2.1
Particle density in water g/cc[c]	—	1.5–1.6	1.5–1.6
Effective size			
Single media filters	0.5 mm	—	0.5 mm
Multi-media filters	0.8 mm	0.8 mm	—
Uniformity coefficient	Less than 1.75	1.9 or less	1.7 or less

[a]Coal based granular activated carbon, products of Calgon Corp.
[b]U.S. Sieve Series.
[c]The pores of the activated carbon filled with water.

Typical size distributions of a commercial carbon are shown in Table 8-2.

TABLE 8-2   TYPICAL SIZE DISTRIBUTION OF GRANULAR ACTIVATED CARBON

	8 × 30 Mesh	14 × 40 Mesh
Sieve Size U.S. Standard Series		
Larger than No. 8 – max %	3	—
Larger than No. 14 – max %	—	3
Smaller than No. 30 – max %	1	—
Smaller than No. 40 – max %	—	1
Mean particle diameter (mm)	1.6	0.9

Specific discussions of the efficiency of activated carbon for filtration are presented in later portions of this chapter.

The head loss through granular carbon is, of course, a function of the carbon size, the depth of the carbon layer, the hydraulic throughput rate, and the water temperature. Figure 8-2 presents the head loss data for some commercial carbons in downflow service as a filter media following backwashing. The backwash characteristics are shown in Figure 8-3.

Other Design Criteria of Concern

Granular carbon can be used after conventional filtration of suspended matter or, of primary interest here, as a combination filtration-adsorption medium. Granular carbon has been utilized in conventional rectangular or circular rapid sand filters, in pressure vessels, and in specially designed adsorbers. The choice of equipment depends upon the severity of the organic removal problem, the availability of existing equipment, and the desired improvement of adsorption conditions.

Usually, 2 or more units are used in parallel downflow operation. The startups of the units are staggered so that the exhaustion of each bed will be in sequence. Blending of the fresh carbon effluent with partially exhausted carbon effluent in effect prolongs the life of the bed before reactivation or replacement of the individual beds is necessary.

Flow rates are usually 2.5–5.0 gpm/ft² and bed depths are normally 2.5–10 ft. Varying the combined values of these two factors adjusts the contact time of the water and the granular carbon beds. A direct linear relationship between contact time and carbon bed performance has been found in full-scale plant tests and concurrent small column tests

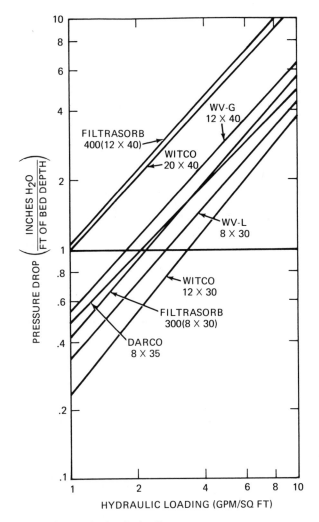

Fig. 8-2 Pressure drop vs hydraulic loading.

(see Table 8-5). Carbon performance at a given contact time has been found to be unaffected by variations in hydraulic loading rates in the 2–10 gpm/ft² range. Thus, in terms of adsorption only, contact time is the governing criteria and not surface loading rate within the ranges of practicality.

When the granular carbon bed is functioning both as a turbidity removal and adsorption unit, there may be reasons to limit the bed depth and flow rate parameters to remove effectively turbidity and to backwash properly the filter. If granular activated carbon is to be effec-

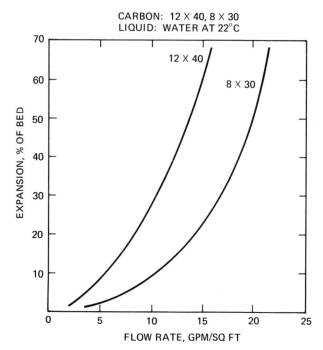

Fig. 8-3 Expansion of carbon bed at various flow rates. Carbon: 12 × 40, 8 × 30; liquid: water at 22°C.

tive in turbidity removal, it must be hard enough to withstand vigorous backwash agitation. At the same time, it should be dense enough to expand during the backwash cycle and to settle quickly for immediate resumption of filtration. As discussed earlier, coal-based granular carbon possesses approximately the same density and filtration characteristics as anthracite, which has found increasingly wider use in the water field.

Particle size of the carbon, in addition to contact time, should be considered carefully as a design factor. Reduction of particle size for a given set of flow conditions is a means of increasing adsorption rates and, thereby, improving adsorption performance. Reduction in particle size to improve adsorption must be consistent with other significant factors such as head loss and backwash expansion. Length of filter run in an adsorption-filtration bed would also be a problem, if too small a particle size were chosen.

Where existing rapid sand filters are being converted to adsorption-filtration units, the permissible depth of the carbon layer will be limited by the freeboard available in the existing structure. Adequate space between the carbon surface and the wastewater troughs should be available to permit at least 30 percent expansion of the carbon layer.

The remainder of this chapter is devoted to discussion of experiences gained at full-scale installations of combined carbon adsorption-filtration units.

EXPERIENCES WITH INTEGRATED FILTRATION-ADSORPTION UNITS

Table 8-3 summarizes the municipal water treatment plants in the United States which utilize granular activated carbon. The remainder of this section will summarize the design and experiences at those installations with a significant operating history.

TABLE 8-3 GRANULAR CARBON INSTALLATIONS IN MUNICIPAL PLANTS IN THE UNITED STATES

Water Plant Location	Year Installed	Size of Plant (mgd)	Flow Rate (gpm/ft²)	Carbon Bed Depth
AWWS Co., Hopewell, Virginia	1961	3.0	2.0	24 in.
Nitro, West Virginia	1966	10.0	1.5–2.0	30 in.
Montecito Co. Water District, Santa Barbara, California	1963	1.5	6	12 ft
Del City, Oklahoma	1967	5.25	2	36 in.
Somerset, Massachusetts	1968	4.5	2	11 in.
Pawtucket, R. I.	1969	24	2	18 in.
Lawrence, Massachusetts	1969	10	2	24 in. 30 in.
Piqua, Ohio	1969	8	2	18 in.
Bartlesville, Oklahoma	1970	4.5	2	24 in.
Granite City, Illinois	1971	7	1.4	18 in.
Winchester, Kentucky	1970	1.5	2	24 in.
Mt. Clemens, Michigan	1968	7	1.7	24 in.

Nitro, West Virginia

The Nitro, West Virginia water treatment plant draws its water from the Kanawha River which is heavily polluted with various organic industrial wastes. Because of a persistent problem with odors, even with the use of double aeration and relatively high powdered carbon and chlorine doses, the use of granular carbon was investigated.[3] Threshold odors of the raw water vary from 300 to 2000. During the initial tests at Nitro, filtration of the plant's 7 mgd output through 2 gravity, rectangular granular-activated carbon filters for 48 days demonstrated the ability of granular carbon to reduce high carbon-chloroform extract (CCE) to within recommended levels (200 ppb). These carbon units were 28 × 28 ft with a 5 ft medium depth, and operated at 3.7 and 4 gpm/ft². 84 to 89.5 percent of CCE contaminants were removed under conditions of high concentration of CCE organics, low-temperature water, and filter rates of approximately 4.0 gpm/ft². These carbon filters were still removing 84 percent of CCE organics at the end of the test period, some 30 days after the filter effluent had reached an unsatisfactory odor value (see Table 8-4).

Carbon column adsorption tests performed during the same period of operation indicated that greater removal efficiency could be obtained by increasing the contact time (Table 8-5). Lowering the flow rate to obtain a contact time of 7.5 min raised the efficiency to 90 percent.

The tests conducted at Nitro also found that a finer 20 × 50 mesh carbon produced an odor-free water for twice as long as the 26 days typically provided by the coarser 8 × 30 mesh carbon. With a contact time of 7.5 min the carbon dosage during winter months when the organic load was relatively high was estimated at 0.7 to 0.13 lb per 1000 gal.

Following the tests, the rapid sand filter media in the 10 plant filters was replaced by a 14 × 40 mesh carbon bed of 30 in depth. The carbon

TABLE 8-4 CCE REMOVAL BY ACTIVATED-CARBON FILTERS AT NITRO, W. VA. (November–December 1963)

Test Period days	CCE		
	Present in Influent ppb	Present in Effluent ppb	Percentage Removed
1–10	2000	210	89.5
11–20	1000	130	87.0
21–30	960	150	84.5
31–40	1200	190	84.0
41–48	1200	190	84.0

TABLE 8-5 INFLUENCE OF CONTACT TIME ON PASSAGE OF SPECIFIC
CONTAMINANTS AT NITRO, W. VA.

	Contact Time — min			
	1.9	3.8	5.6	7.5
Contaminant	Depth — ft			
	5	10	15	20
	Concentration — ppb			
Ethylbenzene	20	18	15	5
Bis (2-chloroethyl) ether	94	44	<D	<D
2-Ethyl hexanol	57	20	<D	<D
Bis (2-chloro-isopropyl ether	26	10	<D	<D
α-Methyl benzyl alcohol	62	13	<D	<D
Acetophenone	11	<D	<D	<D
Isophorone	12	9	<D	<D
Total	285	114	15	5
Threshold odor number (25°C)	32	25	5	2

is supported above the pipe lateral underdrain by coarse gravel in the
conventional concrete gravity filter units. The filters are operated at
throughput rates of 1–2 gpm/ft² and backwashed at 10 gpm/ft². Spent
carbon is removed as a water slurry by a travelling eductor which is
supported from an overhead trolley (see Figure 8-4). The plant includes
carbon regeneration facilities which operate about 25 percent of the
time.

The carbon filters have been in service since 1966. Figures 8-5 and
8-6 and Table 8-6 summarize the performance of the plant. The carbon
beds, which also provide the only filtration, are backwashed about once
every 4 days. The threshold odors, carbon-chloroform extract, and the
turbidity of the finished water have consistently met the USPHS drinking water standards.

Del City, Oklahoma

The Del City, Oklahoma plant design provides adsorption-filtration of a reservoir supply for taste and odor removal.
The plant and the combined adsorption filtration units are illustrated
in Figures 8-7 and 8-8. At a filtration rate of 2 gmp/ft², the filter capacity
is 5.25 mgd. The filter galley is composed of 4 bays with each bay
consisting of 2 split filter beds with a total filter area of 1872 ft². The

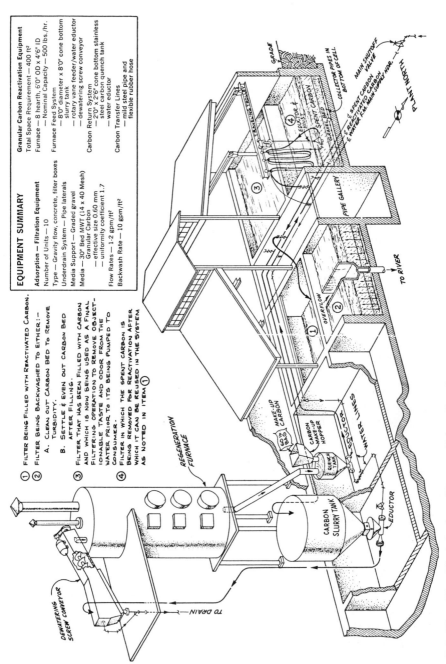

EQUIPMENT SUMMARY

Adsorption — Filtration Equipment
Number of Units — 10
Type — Gravity flow, concrete, filter boxes
Underdrain System — Pipe laterals
Media Support — Graded gravel
Media — 30" Bed MWT (14 x 40 Mesh)
Granular Carbon
  — effective size 0.60 mm
  — uniformity coefficient 1.7
Flow Rates — 1-2 gpm/ft²
Backwash Rate — 10 gpm/ft²

Granular Carbon Reactivation Equipment
Total Space Requirement — 400 ft²
Furnace — 8 hearth, 6'0" OD x 4'6" ID
  — Nominal Capacity — 500 lbs./hr.
Furnace Feed System
  — 8'0" diameter x 8'0" cone bottom
    slurry tank
  — rotary vane feeder/water eductor
  — dewatering screw conveyor
Carbon Return System
  — 2'0" x 2'6" cone bottom stainless
    steel carbon quench tank
  — water eductor
Carbon Transfer Lines
  — mild steel pipe and
    flexible rubber hose

① FILTER BEING FILLED WITH REACTIVATED CARBON.

② FILTER BEING BACKWASHED TO EITHER:—
  A. CLEAN OUT CARBON BED TO REMOVE
     TURBIDITY.
  B. SETTLE & EVEN OUT CARBON BED
     AFTER FILLING.

③ FILTER THAT HAS BEEN FILLED WITH CARBON
   AND WHICH IS NOW BEING USED AS A FINAL
   FILTERING OPERATION TO REMOVE OBJECT-
   IONABLE TASTE AND ODOR FROM THE
   WATER PRIOR TO ITS BEING PUMPED TO
   CONSUMER.

④ FILTER IN WHICH THE SPENT CARBON IS
   BEING REMOVED FOR REACTIVATION AFTER
   WHICH IT CAN BE REUSED IN THE SYSTEM
   AS NOTED IN ITEM ①

Fig. 8-4 Nitro, West Virginia plant. (Courtesy Calgon Corp.)

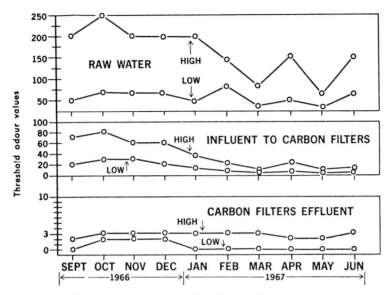

Fig. 8-5 Threshold odor removal at Nitro, West Virginia.[6]

carbon layer (0.8–0.9 mm effective size) is 36 in. deep and overlays a 12 in. deep sand layer (0.5 mm effective size). A gravel bed–Wheeler block underdrain system is provided. Backwashing is accomplished at a rate of 15 gpm/ft² maximum rate. Regeneration facilities are not yet provided at Del City although space has been alloted at the plant for future addition of such facilities. Stainless steel troughs are provided for removal of the carbon in a slurry form from the beds.

Mt. Clemens, Michigan

Taste and odor problems at Mt. Clemens, Michigan have been varied and severe for 33 years. Source of the raw water supply is Lake St. Clair, a shallow lake between lakes Huron and Erie in the chain of the five Great Lakes.

For many years the most intense tastes and odors were confined to the runoff periods when septic and phenol tastes and odors were experienced from pollution by the Clinton River. Industrial development during, and after, World War II increased the pollution problems from the river. By 1958, the phenol had been eliminated from the river system. However, in the summer of that year, the city began to experience intense musty-moldy tastes and odors caused by biological growths in Lake St. Clair.[1]

TABLE 8-6 TURBIDITY REMOVAL AT NITRO, WEST VIRGINIA[6]

		Raw Water	Influent to Filters	Plant Effluent (Less Than)
		Jackson Turbidity Units		
January	maximum	55	3	0.05
	average	32	2	0.05
	minimum	15	2	0.05
February	maximum	200	5	0.05
	average	45	3	0.05
	minimum	15	2	0.05
March	maximum	1300	10	0.05
	average	176	3	0.05
	minimum	15	2	0.05
April	maximum	—	—	0.05
	average	—	—	0.05
	minimum	—	—	0.05
May	maximum	1500	3	0.05
	average	119	2	0.05
	minimum	15	1	0.05
June	maximum	20	3	0.05
	average	19	2	0.05
	minimum	15	1	0.05

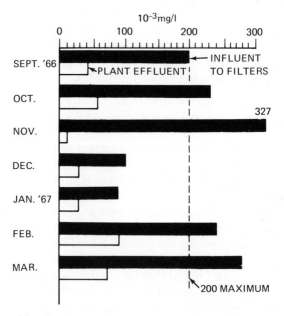

Fig. 8-6 Carbon-chloroform extract removal at Nitro, West Virginia.[6]

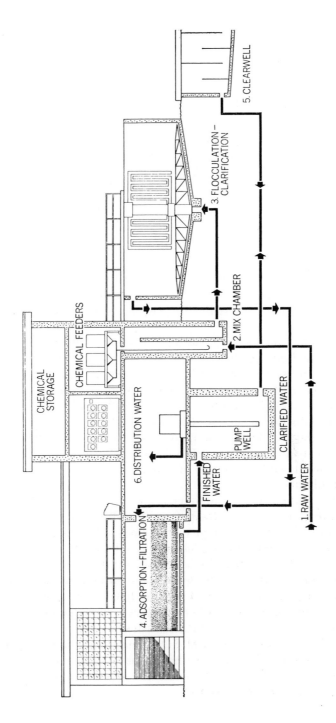

Fig. 8-7 Cross section of Del City, Oklahoma, plant.[5]

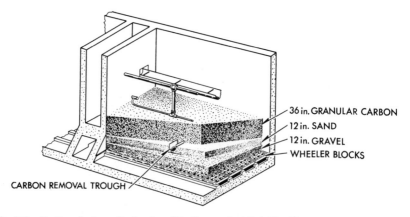

36 in. GRANULAR CARBON
12 in. SAND
12 in. GRAVEL
WHEELER BLOCKS

CARBON REMOVAL TROUGH

Fig. 8-8   Sectional view of adsorber-filtration unit at Del City.[13]

Dry, powdered activated-carbon treatment was capable of removing these tastes and odors, but the city's purification department could not apply enough of it with dry feed machines to have it always under control.

In 1967, the decision was made to convert the existing sand filters to combined adsorption-filtration units by replacing the upper 24 in. of the 30 in. sand filters with granular carbon. Carbon with an effective size of 0.55–0.65 mm and a uniformity coefficient of 1.4–1.7 was placed over the 2832 ft² of filter area in the 8 plant filters. Filter revision was accomplished in the February–April, 1968 period. Some carbon fines appeared in the filter effluent until June, 1968 but none have been noted since. The filter performance has been reported to be better than for the all-sand filters used previously. There have been no taste or odor complaints.

The carbon was found to have a useful life of 3 years — equivalent to about 2.7 mg of water treated per ft² of filter area (2 ft³ of carbon). This is equivalent to an average carbon dose of 2.6 mg/l and a carbon cost of $4.57/mg. The cost to replace the spent carbon was found to be about $0.51/mg (0.07 hr/mg of labor) based on the above throughput. Carbon was removed by shovelling the spent carbon into an eductor.

Filter rates as high as 4 gpm/ft² have been used. Initial operating problems were reported to be (1) loss of some carbon due to excessive pressure on the Palmer surface wash arms which was solved by reducing the pressure to 50 psi, and (2) some channeling through the filters if they were run to excessive head loss — they are now backwashed at 4½–5 ft of head loss. The plant is now reported to be providing satisfactory results in every respect.

Hopewell, Virginia

At Hopewell, filters with fresh carbon were placed in service along with similar sand medium filters. The carbon beds were 24 in. deep and were tested for both adsorption and filtration at conventional sand filter rates. For 60 days, the carbon filters reduced threshold odor from 70 to 4. At the same time, they reduced turbidity to less than 0.07 Jackson Units — a performance somewhat superior to that of the sand filters. Superchlorination preceded the filtration, and free residual chlorine was reduced from 1.4–2.8 ppm to less than 2.25 ppm. After their odor removal capacity was exhausted, the carbon filters continued to produce water that, in regard to its color and to iron, manganese, chlorine, and turbidity content, was of a quality equivalent to or better than that produced by the sand filters.

Other Projects

It has recently been reported that EPA is evaluating plants at Lawrence, Massachusetts and Davenport, Iowa which have installed granular carbon filters.[12] Preliminary results at the Lawrence plant showed that the carbon reduced the influent Threshold Odor Number from 8–9 to 3 or less for a period of 80–90 weeks. However, in only 16 weeks, the CCE removals reached zero. The CAE removals were found to be zero from the start. The EPA tests at Davenport will include similar measurements to determine if a similar pattern occurs.

GRANULAR CARBON REGENERATION

Techniques for granular carbon regeneration are discussed in detail elsewhere.[8] Basically, the adsorbed organics are driven from the granular carbon by heating the carbon to temperatures above 1500°F in a multiple hearth furnace. Spent carbon is typically removed from the carbon bed in a slurry, dewatered by gravity drainage in a storage vessel, fed into the furnace where the regeneration occurs, water quenched as it leaves the furnace, and is then stored or returned to the carbon filter (see Figure 8-4). Carbon losses of 5 percent per regeneration cycle can be maintained.

Whether or not carbon regeneration is practical is dependent upon the plant capacity and the carbon dosages required to achieve the desired water quality. The smallest commercial furnaces have capacities of 2500 to 3500 lb per day (910,000–1,280,000 lb per year if operated continuously). Table 8-7 summarizes the reactivation requirements for various carbon dosages for a 10 mgd plant. As tertiary treatment of

TABLE 8-7   REGENERATION REQUIREMENTS FOR A 10 MGD PLANT AT VARIOUS CARBON DOSAGES

| Carbon Dosage | | Carbon Regenerated |
lb/mg	mg/l	Per Year, lb
38	4.6	140,000
58	7	210,000
115	14	420,000
330	28	840,000

secondary sewage effluents usually requires only 250–350 lb per million gal, it is apparent that the upper limit of this table represents very severe operating conditions. It is also apparent that even a small regeneration furnace is adequate for most 10–20 mgd water treatment plants. The costs of such a regeneration system will vary from $50,000–$90,000. Reactivation costs vary substantially depending on how frequently and for what operating periods the furnace may be run. For example, regeneration of 140,000 lb/year of carbon in a 2500 lb/day capacity system may result in costs (capital and operating) of 7–8¢/lb of carbon while the same system might regenerate 840,000 lb/year of carbon at a cost of 4¢/lb of carbon. Virgin carbon costs are about $0.30/lb. The cost of carbon treatment on such a basis in combined adsorption-filtration units is in the range of $3 to $13/mg. For most raw waters, carbon dosages lower than those shown in Table 8-7 will be adequate. As noted earlier, the average dosage required at Mt. Clemens, Michigan was only 2.6 mg/l over a 3 year period.

For cases where very low carbon dosages — such as Mt. Clemens — are adequate, single use of the carbon may be economically acceptable. For example, the Mt. Clemens costs were only about $5/mg on a single use basis. Shipping spent carbon to a central regeneration facility may be practical in some areas.

Granular carbon has several advantages over powdered carbon — cleanliness, ease of handling, and self-adjustment to changing raw water characteristics. On-site regeneration, in some cases, will enable these advantages to be realized at very reasonable costs while single use of carbon may be practical where low carbon dosages are required.

CARBON TRANSPORT

The primary use of air or pneumatic transport of carbon is in bulk-handling of make-up carbon. Once carbon is introduced into the adsorption-regeneration system, it is usually transported hydraulically in slurry form.

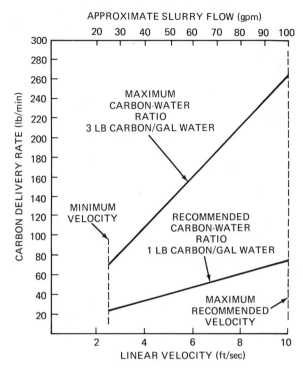

APPROXIMATE SLURRY FLOW (gpm)

Fig. 8-9 Carbon delivery rate (2 in. pipe).

Handling characteristics have been experimentally studied by using water slurries of 12 × 40 mesh granular carbon in a 2 in. pipeline. The data indicate that a maximum of 3 lb of carbon per gal of water could be transported hydraulically but that it is better to use 1 gal of water for moving each lb of carbon. The velocity necessary to prevent settling of carbon is a function of pipe diameter, granule size, and liquid and particle density. The minimum linear velocity to prevent carbon settling was found to be 3.0 ft/sec. It is recommended that a linear velocity between 3.5 to 5.0 ft/sec be used. Velocities of over 10 ft/sec are objectionable due to carbon abrasion and pipe erosion. Carbon delivery rates are a function of pipe diameter, slurry concentration, and linear velocity. Data are shown in Figures 8-9 and 8-10. Pressure drop data for various slurry concentrations velocities in 2 in. pipe are shown in Figure 8-11.

Pilot plant tests indicate that after an initial higher rate, the rate of attrition for activated carbon in moving water slurries is approximately constant for any given velocity, reaching an approximate value of 0.12 percent fines generated per exhaustion-regeneration cycle. This de-

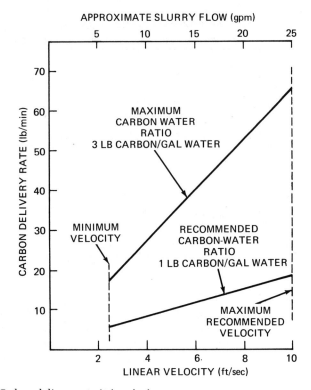

Fig. 8-10   Carbon delivery rate (1 in. pipe).

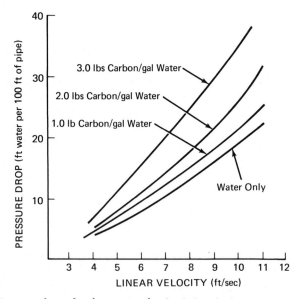

Fig. 8-11   Pressure drop of carbon-water slurries (2 in. pipe).

terioration of the carbon with cyclic operation has been reported to be independent of the velocity of the slurry (within the range recommended previously — 3.5 to 5 ft/sec). Loss of carbon by attrition in hydraulic handling apparently is not related to the type of pump (diaphragm or centrifugal) used.

Carbon slurries can be transported by using water or air pressure (blowcase), centrifugal pumps, eductors, or diaphragm pumps. The choice of motive power is a combination of owner preference, turndown capabilities, economics, and differential head requirements.

ACTIVATED CARBON FOR DECHLORINATION OF WATERS

Carbon has been used for over half a century to dechlorinate water, the early systems using granular coal and coke. Hans Pick was the first to evaluate the parameters of the dechlorination process. The proposed mechanism was:

$$2Cl_2 + C + 2H_2O \rightleftarrows 4HCl + CO_2$$

However, this reaction describes the summation of the reactions taking place on the surface of the carbon. The actual reaction according to Kovach[11] is:

$$Cl_2 + H_2O \rightarrow 2H^+ + 2Cl^- + O$$

The chemisorbed nascent oxygen (last term on right) decomposes in either of the following two ways:

$$C_xO_x \rightarrow C + CO$$
$$C_xO_x \rightarrow C + CO_2$$

Pick also introduced an empirically derived equation to describe the process:

$$\log\frac{C_o}{C} = k\,\frac{L^{1/2}}{V}$$

where C_o is inlet chlorine concentration: C, outlet chlorine concentration; L, bed length; k, a constant; and V, flow rate.

The process of chlorine removal from water is not a pure adsorption process but also involves a chemical reaction between the chlorine and the water. The initial step is the adsorption of chlorine in the form of hypochlorous acid on the carbon surface. The subsequent decomposition of the hypochlorous acid results in the formation of hydrochloric acid and nascent oxygen.

The hydrochloric acid is not readily adsorbed and is released into solution, while the chemisorbed oxygen builds up to form an oxide complex on the carbon surface. This surface oxide, which reduces the efficiency of the dechlorination process, is fairly stable at low reaction rates. At high inlet chlorine concentrations (1000 ppm), free carbon monoxide and carbon dioxide can be found in the effluent water. Removal of the surface oxide is incomplete below 400°C, making *in situ* regeneration uneconomical. It is postulated that even under conditions of industrial dechlorination (below 30 ppm chlorine), some carbon dioxide is generated; however, the quantity is too low to be detectable against the background carbon dioxide content. Several investigators also postulated the formation of $HClO_3$ above 2660 ppm chlorine concentration, but recent work shows that the thermodynamics of this reaction, at least in static systems, is unfavorable.

The basic design parameters related to the operation of an activated carbon dechlorinator are: (1) chemical state of the free chlorine, including pH, (2) inlet chlorine concentration, (3) flow rate through the carbon bed, (4) particle size of the carbon, (5) type of carbon, (6) temperature, (7) manner of operation (continuous or interrupted), and (8) impurities present.

It has also been proposed that the dechlorination process is controlled by the slow chemical reaction on the surface; i.e., decomposition of HOCl, in which case Pick's equation above can be modified to:

$$\log \frac{C_0}{C} = k_1 \frac{L}{V}$$

Based on current data, the use of either equation can be justified depending on the conditions of the experiment. Due to the fact that k_1 is less than k, the latter equation contains a built in safety factor.

While the establishment of a design based on either equation can be performed easily, the expected variations in process conditions are rarely taken into account. The equations show the importance of using the actual chlorine concentrations expected in the influent water; acceleration of the test by using artificially high chlorine concentrations for rapid carbon evaluation leads to faulty conclusions.

The chemical state of the free chlorine is one of the strong influencing factors of the dechlorination process. For example, chloramines considerably reduce the dechlorination efficiency. Unfortunately, no quantitative evaluation was available to the authors on the effects of chloramines. In some instances where chlorine dioxide is used, the effect of pH becomes important because the $NaClO_2$ formed is more stable than HOCl; thus, any evaluation of activated carbon should be subjected to the same chlorine forms, or the pH should be adjusted.

Dechlorination efficiency for any activated carbon is also strongly dependent on particle size, and the smallest mean particle size should be selected for any carbon grade. Of course, the limiting factor in such a selection is the maximum allowable pressure drop. The micro- and macroporosity of the carbon also influence dechlorination efficiency. Since the dechlorination process is related to the surface area of the carbon, macropores are necessary to make the high surface areas accessible for the adsorption reaction. The raw material from which the carbon is made generally does not have any affect on dechlorination efficiency.

The presence of colloidal impurities or high organic concentrations can substantially reduce the carbon useful life as a dechlorinator. The colloids may plug the pores of the carbon, reducing the surface area available for dechlorination. Organics can form an adsorbate film on the carbon, causing the same end result. If pretreatment steps are adequate to remove these impurities, beds of activated carbon will have a long useful life as a dechlorinating bed.

If the experimental conditions are carefully selected so that they match the expected operating conditions, then it is fairly straightforward to evaluate activate carbon in dechlorination and to size the adsorber.

Recent tests have determined efficiency values for specific carbons available to industrial and municipal treatment operations in the United States. The results for flow rates, concentration of influent, and type of carbon are shown in Figure 8-12. They are based on a chlorine breakpoint of 0.01 ppm. The pH of the water was 7 and the temperature was 23°C. The absence of bacteria or any organic interference were assumed.

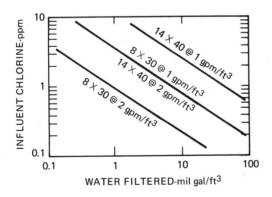

Fig. 8-12 Effect of flow rate and mesh size on dechlorination by granular carbon filters.

The life of the carbon in dechlorination service is extremely long. For example, under conditions of sand filter service (that is, 2.5 gpm/ft^2 and 2.5 ft bed depth), granular carbon medium in a 1 mgd filter (700 ft^3) on dechlorination service alone could process 700 million gal of 4 ppm

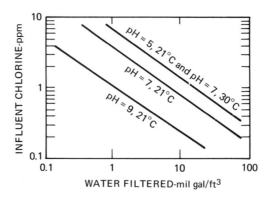

Fig. 8-13 Effect of *p*H and temperature on dechlorination by granular carbon (8 × 30 mesh at 1 gpm/ft³).

free-chlorine influent water before a breakpoint of 0.01 ppm chlorine would be reached. A bed processing water containing 2 ppm chlorine under similar conditions would last about 6 years. The effect of mesh size is pronounced. As indicated in Figure 8-12, a reduction in particle size, reflected in the reduction of mesh size from 8 × 30 to 14 × 40, allows a doubling of flow rate without a sacrifice in efficiency.

Dechlorination will proceed concurrently with adsorption of organic contaminants. Long-chain organic molecules, such as those of detergents, will reduce dechlorination efficiency somewhat, but many common water impurities, such as phenol, have little apparent effect upon the dechlorination reaction.

A rise in temperature and a lowering of *p*H favor dechlorination. Figure 8-13 indicates the relationship of these factors as they vary from *p*H 7 and 21°C. Mesh size was 8 × 30 and flow rate was 1 gpm/ft³. A breakpoint of 0.01 ppm Cl_2 and the absence of bacteria or any organic interference were assumed. It is unlikely that a deliberate change in *p*H or temperature favoring dechlorination alone would be economically feasible, unless existing conditions significantly retard the process. These data are values determined with chlorine in distilled water. Variance in hydraulic loading, suspended matter, and certain adsorbed organics, as noted above, could adversely affect dechlorination efficiency.

REFERENCES

1. Hager, D. C., and Flentje, M. E., "Removal of Organic Contaminants by Granular Carbon Filtration, *Journal American Water Works Association*, **57**, p. 1440 (November 1965).

2. Hansen, R. E., "Granular Carbon Filters for Taste and Odor Removal," *Journal American Water Works Association,* **57,** p. 176 (March 1972).

3. Dostal, K. A., Pierson, R. C.; Hager, D. G.; and Robeck, C. C., "Carbon Bed Design Criteria Study at Nitro, W. Va.," *Journal American Water Works Association,* **57,** p. 663 (May 1965).

4. Flentje, M. E., and Hager, D. C., "Advances in Taste and Odor Removal with Granular Carbon Filters," *Water and Sewage Works* (February 1964).

5. Klaffke, K., "Granular Carbon System Purifies Water at Del City, Oklahoma," *Water and Wastes Engineering* (March 1968).

6. Hager, D. C., and Fulker, R. D., "Adsorption and Filtration with Granular Activated Carbon," *Journal of the Society for Water Treatment and Examination,* **17,** p. 41 (1968).

7. "Water and Waste Treatment with Filtrasorb Granular Activated Carbon," Calgon Corp., 1971.

8. Culp, R. F., and Culp, G. L., *Advanced Wastewater Treatment* Van Nostrand Reinhold Co., New York, 1971.

9. "Activated Carbon and Wastewater," Westvaco, 1971.

10. Hyndshaw, A. Y., "The Selection of Granular Versus Powdered Activated Carbon," *Water and Wastes Engineering,* p. 49 (February 1970).

11. Kovach, J. L., "Activated Carbon Dechlorination," *Industrial Water Engineering,* p. 30 (Oct.–Nov., 1971).

12. Symons, J. M., and Robeck, G. C., "Treatment Processes for Coping with Variations in Raw Water Quality," presented at the 1973 American Water Works Association Conference, Las Vegas, Nevada, May 16, 1973.

9 Wastewater in Supply Sources

INTRODUCTION

Ever since the water-carried waste systems were first introduced over 100 years ago, concerns have been expressed about the possible effects on downstream water supplies and the health of water drinkers. Water filtration and chlorination have completely overcome the problems and concern about transmission of waterborne bacterial diseases such as typhoid fever, dysentery, and the like, when these processes are properly applied and operated. As increasing quantities of wastewater and a wide variety of new contaminants have been introduced to water supply streams, professionals and laymen alike are increasingly concerned about possible new threats to health and safety from discharges above water intakes. Some of the categories of principal concern at this time are viruses, increases in total dissolved solids, trace organics, heavy metals, and biostimulants. In the past, a general practice was to utilize insofar as possible the assimilative capacity of the receiving stream to achieve maximum economy of wastewater treatment. Current trends are toward applying the best treatment technology available, to provide maximum and highest use of water resources. From comparison of conventional and advanced wastewater treatment (AWT) results, a tremendous improvement in water quality and safety

256

in water reuse is obviously achieved by application of the latest techniques. In many situations, wastewater can be treated to exceed the quality of the receiving stream in most, if not all, respects.

CONCEPTS OF WATER REUSE

The indirect reuse of water is a very old, widespread, and
generally accepted practice in the public water supply industry of the
United States. For the most part the reuse is unplanned and uninten-
tional. It results from the discharge to streams of partially treated or
untreated wastewaters followed by the downstream use of the co-
mingled runoff and wastewater in the stream as a source of municipal
water supply. Over the past 50 years, this type of water reuse has
proved to be a satisfactory practice despite the much greater pollutional loads and increased water withdrawals developed during this
time.

The quality and safety of public water supplies in the United States
are considered to be the best in the world. This opinion is well-
supported by public health statistics and epidemiological evidence.
Waterborne disease has virtually been eliminated in this country. This
favorable situation is due primarily to the good design, operation, and
control of water purification plants and distribution systems. It is also
due to treatment of wastewaters by the combination of plant processes
and natural phenomena. This type of water reuse, which utilizes both
natural purification and plant processing, is referred to as indirect
reuse.

Indirect reuse may be either uncontrolled or controlled. Uncontrolled indirect reuse has been generally described above, and more
specific details about it will be presented later. An example of controlled indirect reuse is the intentional artificial recharge of groundwater
aquifers by adequately treated wastewaters, which currently is also a
well-accepted practice. Further examples are given later. Controlled
indirect reuse should include not only the natural safeguards of dilution and separation in time and space which apply to some degree in
uncontrolled reuse, but also the most recent developments in AWT.
The addition of AWT techniques to conventional treatment processes
greatly reduces the pollutional load to be assimilated by the receiving
waters, which in turn increases the efficiency and reliability of the
natural purification processes. Even more importantly, the use of advanced treatment methods prior to discharge of wastewaters also provides for the removal of a wide variety of objectionable substances
which are only partially removed or completely unaffected by natural
processes. The introduction of advanced treatment into the indirect
reuse cycle tremendously increases the safety of such reuse.

Controlled indirect reuse must not be confused with direct potable reuse. Direct potable reuse is defined as the reuse which results from the deliberate treatment and conveyance of wastewater through man-made facilities directly to a point of mixing with natural waters at the intake of a public or domestic water system. Direct potable reuse systems are often described as "closed loop" or "pipe-to-pipe" systems. At the present time, direct potable reuse is not a necessity, and is not proposed. It is important to distinguish between controlled indirect reuse and direct potable reuse because the terms are often loosely used, even in the technical literature.

Uncontrolled Indirect Reuse

A study by the Federal Water Pollution Control Administration, in which 155 cities using surface water supplies were sampled, provides good insight into the extent of uncontrolled indirect reuse in the United States. Upstream municipal wastewaters present in the water supply sources during low flow months ranged from 0 to over 18 percent, with a median, based on population served, of 3.5 percent. This means that 1 gal out of every 30 has previously passed through the wastewater system of upstream communities.

At Jefferson City, Mo. the reuse factor was 5.7 percent for low flow months and 1.8 percent at average flow. At Kansas City, the figures are 6.6 and 2.4 percent, respectively, and at Omaha are 2.3 and 1.0 percent. The figures include only municipal wastewater flows and not industrial discharges.

Other data on return flows indicate that the total municipal and industrial wastewater, exclusive of cooling water, discharged to the Missouri River and its tributaries between Gavins Point Dam (northeast Nebraska) and a point above the St. Louis metropolitan area is about 800 mgd. This corresponds to about 1.6 percent of the average flow at the Hermann, Missouri gage located above St. Louis. The corresponding value for the minimum flow month is 5.6 percent.

In Kansas, extensive reuse occurs along the Neosho and Verdigris rivers. Twelve cities use the Neosho as a source of water supply and as the receiving stream for their treated wastewaters. At times in the past, stream flow was zero between the city's water intake and its wastewater outfall, and only occurred between the outfall and a downstream neighbor's waterworks intake. On the Verdigris River, the total amount of water used by Kansas municipalities at times has been about 3 times the river flow at the Oklahoma state line.

An example of uncontrolled indirect reuse is offered by the Occoquan Reservoir which provides the source of drinking water supply for about one half million people in the Washington D.C. area and is filled from the 600 square mile Occoquan watershed. The 1700 acre reservoir has a capacity of 9.8 billion gal of water and can provide up to 65 mgd of drinking water to the Fairfax County Water Authority. Bull Run and Occoquan Creek are tributary to the reservoir. Observed minimum flows are 0.1 and 2.3 cfs, respectively. Discharging within the Occoquan watershed are 25 secondary sewage treatment plants which have a total rated capacity of 5.4 mgd. Many of the plants are operating at or above their rated capacity.

Water produced from the Occoquan Reservoir has been consistently safe for potable use, but the partially treated wastewaters have contributed to the extensive pollution of the reservoir, creating massive blooms of troublesome blue-green algae in the summer and fall which cause objectionable taste and odor in the drinking water supply. In 1969, the water in the reservoir was found to be devoid of dissolved oxygen from 10 ft below the surface to the bottom of the reservoir. An estimated 75 percent of the phosphorus and 40 percent of the nitrogen, the two major algal nutrients in the reservoir, originated from partially treated municipal wastewaters.

Wastewater flow will make up a significant portion of the total water resource offered by the Occoquan reservoir, as the population of the area increases. A report by the Virginia State Water Control Board indicates that a future 40 mgd wastewater flow will make up 12 percent of the total annual inflow and 46 percent of the July low flow to the reservoir.

Projections of the water demands to be served by the Fairfax County Water Authority indicate that the currently available reservoir supply will not be adequate in quantity by 1975. Thus, any diversion of wastewaters out of the drainage basin would aggravate a potentially serious water supply problem. Continued discharge of wastewaters within the watershed requires a high degree of treatment to protect downstream water users. The past and current use of the Occoquan Reservoir offers an example of uncontrolled, indirect use.

Controlled, Indirect Reuse

Occoquan Basin. In order to eliminate the present problems with algal growths in Occoquan Reservoir and the resulting taste and odor in the water supply produced therefrom, the Upper Occoquan Sewage Authority has developed a plan to replace the multiplicity of existing conventional secondary wastewater treatment plants with a single regional water reclamation plant. This plant will include advanced treatment processes which will produce water of an extremely high quality. The plant will be designed to be fail-safe. Each process and each piece of equipment will have redundant or standby elements, so that treatment can continue at full efficiency during equipment and power failures as well as during shutdowns for routine maintenance and repair. The reclaimed water will be closely monitored and then discharged into a holding reservoir at the plant site. This clear water reservoir will provide a recreation facility for the area, provide observation and sampling basin prior to final discharge, and demonstrate the eutrophication control potential of the reclaimed water. The water plant will discharge at a point approximately 20 miles upstream from the Occoquan water supply reservoir, so that water will be conserved for reuse after exposure to natural purification processes in the effluent reservoir and in the Occoquan Reservoir itself.

The Virginia State Water Control Board has set high quality standards for the reclaimed water as shown below (measured on a weekly average basis):

BOD	1.0 mg/l
COD	10.0 mg/l
Suspended solids	0
Nitrogen	1 mg/l of unoxidized N
Phosphorus	0.1 mg/l as P
MBAS	0.1 mg/l
Turbidity	0.4 JTU
Coliform, mpn	<2/100 ml

The treatment process selected to meet these standards is tabulated below:

Liquid Processing

1. Primary setting.
2. Completely mixed activated sludge.
3. Lime coagulation and settling for phosphorus removal.
4. Two stage recarbonation with intermediate settling.
5. Mixed-media filtration.
6. Selective ion exchange (ammonia removal).
7 Chlorination for trace ammonia removal.
8. Granular carbon adsorption.
9. Chlorination.
10. Storage, on site.

Solids Handling

1. Sludge incineration.
2. Lime recovery.
3. Carbon regeneration.
4. Ion-exchange regenerant recovery.

The initial capacity of the plant is 22.5 mgd, with an anticipated ultimate capacity of 40 mgd. In the future, the reclaimed water will make up 12 percent of the total annual inflow and 46 percent of the July inflow to Occoquan Reservoir. The raw water quality will be greatly improved over present conditions of secondary effluent discharge, and the factor of safety from a public health standpoint will be much greater than at present, both from the standpoint of recreational use of the water in the streams and reservoir, and in consumption of the drinking water supply.

The Virginia Water Control Board chose water reclamation for controlled indirect reuse over the alternate of exportation and dilution for many reasons. Secondary treatment and export to the Potomac River would solve the sewage-caused water quality problem in the Occoquan, but it would be very costly and would merely move the problem to the Potomac estuary. Retention of the highly treated reclaimed water within the Occoquan watershed conserves 10 to 40 mgd of excellent quality water for reuse in an area which needs additional water. The planned Occoquan program illustrates controlled indirect reuse as contrasted to the past practice of uncontrolled direct reuse practiced in the same basin.

Tahoe-Indian Creek. The wastewater reclamation plant at South Lake Tahoe, California was the first plant scale advanced wastewater treatment plant to be built in the United States. The present 7.5 mgd plant was placed in operation in April, 1968. It has been in continuous uninterrupted operation since that date. The reclaimed water has always complied with the high standards set by the California State Water Resources Control Board. This plant was the first to use many new processes for wastewater treatment including chemical treatment for phosphorus removal, high-rate mixed-media filtration, and granular carbon adsorption of organics. It also pioneered carbon regeneration and lime recovery and reuse. Nitrogen removal was not a specific requirement, but the Tahoe plant also included an ammonia stripping tower which has been operated on an experimental basis. Some temperature limitations of this process (it is not practical to operate at ambient air temperatures below freezing) and the need to provide means for prevention or removal of calcium carbonate scale have been demonstrated. Based on this experience modifications have been made in more recent tower designs.

The Tahoe effluent, due to state laws which prohibit discharge of any effluents (regardless of quality) in the Tahoe Basin, is exported to form the Indian Creek Reservoir. This 160 acre reservoir is nearly all AWT effluent and has a volume of about 1 billion gal. The Regional Water Quality Control Board has formally approved and authorized use of the reservoir for unrestricted recreational purposes including fishing, boating, water skiing, swimming, and other water contact sports. This approval was granted after 9 months of observation of the AWT plant performance, water testing, and observation of the reservoir, and after consultation and receipt of recommendations from other state agencies, including the California Department of Health. This clear water reservoir also supports an excellent trout fishery, and its water is used for irrigation of pasture, hay, and alfalfa.

In 1971 the average effluent quality at South Tahoe was:

BOD, mg/l	0.7
COD, mg/l	9.0
SS, mg/l	0
Phosphorus, mg/l	0.06
MBAS, mg/l	0.15
Turbidity, JTU	0.6
Coliform, mpn/100 ml	less than 2.0

Two summers of virus testing at Tahoe detected viruses in the secondary effluent but failed to detect viruses in any of the samples of chlorinated reclaimed water.

Orange County (California). A 15 mgd reclamation plant is now under contruction at Orange County, California. The plant will receive trickling filter (secondary) effluent from the Orange County Sanitation District's plant as its source of supply. The water is to be treated by lime coagulation for phosphorus removal, nitrogen removal (by stripping), two-stage recarbonation, filtration, granular carbon adsorption, and breakpoint chlorination. Carbon and lime will be thermally recovered and reused. The reclaimed wastewater will be mixed with an equal volume of desalted sea water, for reduction in total dissolved solids, and injected into the groundwater aquifer to create a barrier against sea water intrusion into the groundwater basin. After dilution and exposure to natural purification processes within the aquifer, the water will be pumped from wells for reuse including potable use. This is a good example of controlled indirect reuse of water.

At Orange County the regulatory agency requirements for constituents in the injection water are as follows:

Constituent	Concentration mg/l (except electrical conductivity and pH)
Electrical conductivity	900 umhos/cm
pH	6.5 to 8.0
Ammonium	1.0
Sodium	110
Total hardness (as $CaCO_3$)	220
Sulfate	125
Chloride	120
Total nitrogen (as N)	10
Fluoride	0.8
Boron	0.5

Constituent	Concentration mg/l (except electrical conductivity and pH)
MBAS	0.5
Hexavalent chromium	0.5
Cadmium	0.01
Selenium	0.01
Phenol	0.001
Copper	1.0
Lead	0.05
Mercury	0.005
Arsenic	0.05
Iron	0.3
Manganese	0.05
Barium	1.0
Silver	0.05
Cyanide	0.2

The injection water shall not cause taste, odors, foam, or color in goundwater.

The filter effluent turbidity shall not exceed 1.0 Unit. The carbon adsorption column effluent shall not exceed a chemical oxygen demand concentration of 30 mg/l.

The chlorine contact basin effluent shall always contain a free chlorine residual.

Nassau County, New York. An experimental groundwater recharge project is in operation at Bay Park, New York. The purpose of the project is to develop information which will help ultimately to develop a line of recharge wells along the south shore of Nassau County to return 30 mgd or more of treated wastewater to the groundwater aquifer. Greater withdrawals from wells for water supply purposes would be permitted, and rate of salt water encroachment retarded. Here 1 mgd of secondary (activated sludge) effluent is treated by alum coagulation, filtration, granular carbon adsorption, and disinfection with chlorine. A summary of operating data is given below:

NASSAU COUNTY, NEW YORK WATER RENOVATION PLANT
SUMMARY OF OPERATING RESULTS

	Secondary Treatment Plant Effluent			Advanced Waste Treatment Plant Effluent		
	Low	Median	High	Low	Median	High
Turbidity, JTU	10.0	40.0	100.0	0.1	0.5	3.0
MBAS, mg/l	1.0	2.0	5.0	0.1	0.1	0.5

	Secondary Treatment Plant Effluent			Advanced Waste Treatment Plant Effluent		
	Low	Median	High	Low	Median	High
PO₄, mg/ l	10.0	25.0	50.0	0.5	2.0	5.0
COD, mg/ l	40.0	60.0	100.0	4.0	5.0	10.0
TOC, mg/ l	20.0	25.0	40.0	2.0	3.0	5.0
pH	6.8	7.2	6.8	5.8	6.5	6.8

This is still another example of satisfactory, controlled indirect reuse.

Windhoek, South Africa. The Windhoek project is being mentioned under controlled indirect reuse even though it might better be classed as direct potable reuse since reclaimed water is taken directly from maturation (or holding) ponds to the intake of the public water system and comprises as much as one third of the total municipal water supply. The secondary effluent is reclaimed for reuse by chemical (lime) treatment, flotation, ammonia stripping, recarbonation, filtration, foam fractionation, and granular carbon adsorption. This reuse for drinking water supply has been in operation for more than 2 years. In the spring of 1972, above normal precipitation completely filled the Windhoek water supply reservoirs and the reclaimed water was not needed. This lull in operations was utilized to make additions and improvements to the reclamation plant.

SPECIFIC CONCERNS RELATED TO WATER SUPPLIES CONTAINING WASTEWATERS

Virus Removal

One of the areas of major concern about the safety of water reuse for purposes involving human contact or ingestion is that of possible transmission of viral disease. A review of the water literature reveals many important considerations concerning designs to insure that plants provide complete virus inactivation or removal.

In a 20 month study of the removal of polio virus Type I from water, Robeck, Clarke and Dostal made several noteworthy observations.[1] Their work disclosed clearly that "coagulation and filtration are really one process and must be studied together." This is now a widely accepted and applied principle. They further state, "Through smaller than coliform organisms, the polio virus organism is removed by flocculation and filtration with about the same efficiency as coliforms." They note that, "If a low but well-mixed dose of alum was fed just ahead of the filters operated at 6 or 2 gpm/sf, more then 98 percent of the viruses were removed by 16 in. of coarse coal on top of 8 in. of sand.

If the alum dose was increased and conventional flocculators and set-tling were used, the removal was increased to over 99 percent." In a paper,[2] "High Quality Water Production and Viral Disease," H. E. Hud-son, Jr. reached the following conclusions:

Filtration plants operated to attain a high degree of removal of one impu-rity tend to accomplish high removals of other suspended materials. Exam-ples of parallelism in removal of turbidity, manganese, microorganisms, and bacteria are cited.

"Speed and simplicity make the turbidity measurement a valuable index of removal of other materials. Plants producing very clear water also tend to secure low bacterial counts accompanied by low incidence of viral diseases.

"The production of high-quality water requires striving toward high goals as measured by several . . . quality criteria. These criteria include filtered-water turbidity, bacteria as indicated by plate counts and by presumptive and confirmed coliform determination.

The operating data for plants treating polluted water indicate that low virus disease rates occur in cities where the water treatment operations aim to produce a superior product rather than a tolerable water.

An AWWA Commitee report[3] by Clarke, Berg, Liu, Metcalf, Sullivan, and Vlassoff summarizes the status of viruses in water. Excerpts from this report follow:

Any virus excreted and capable of causing infection when ingested could be transmitted by water. Practical conditions would seem, however, to indi-cate that we should be concerned with only those viruses that can multiply in the intestinal wall and that are discharged in large numbers in feces. This means that the recognized viruses with which we should be concerned are the members of the enteric virus group: polioviruses, Coxsackie viruses, ECHO viruses, the virus (es) of infectious hepatitis, the adenoviruses, and the reoviruses. The total number of virus types that compose these 6 major groups is around 100 and, with the exception of the agent (s) of infectious hepatitis, they have been shown to be present in sewage. Isolation of hepatitis viruses has not been possible to date.

What can be said about gastroenteritis and the related diarrheal diseases?

Many epidemics have been conveniently ascribed to viruses, although very, very few have been thoroughly studied to identify causative agents. Experi-mental work does, however, indicate that some gastroenteritis or diarrheal disease is caused by viruses and that, under the proper circumstances, could be waterborne.

A question repeatedly asked those involved in research on the problem of viruses in water, and probably those involved in supplying the public with drinking water, is "Is water that meets USPHS bacteriological drinking water standards free from disease-producing levels of virus?" This question can perhaps be answered if the following factors are considered:

● What are the relative numbers of indicator bacteria and viruses in water and sewage?
● What is the relative survival time of the two groups of microorganisms?

- What is the relative effectiveness of water and sewage treatment processes on indicator bacteria and viruses?
- What epidemiologic evidence do we have that water meets or does not meet the standards can be responsible for virus disease or infection in humans?

Relative numbers of coliform bacteria and enteric viruses in water and sewage have been discussed by Clarke, et al. These authors have calculated coliform–virus ratios to be about 92,000:1 in sewage and about 50,000:1 in polluted surface waters. Coliform organisms greatly outnumber enteric viruses in sewage or polluted surface waters and therefore appear to be a far better indicator of pollution than enteric viruses.

There is no doubt that water can be treated so that it is always free from infectious microorganisms — it will be biologically safe. Adequate treatment means clarification (coagulation, sedimentation, and filtration), followed by effective disinfection. Effective disinfection can be carried out only on water free from suspended material. The importance of this latter point has been clearly indicated by Sanderson and Kelly. They describe a situation in which coliforms were consistently isolated from waters containing from 0.1–0.5 mg/l free chlorine and between 0.7–1.0 mg/l total chlorine after 30 minutes contact time. This water had turbidity values of from 3.8 to 84 Units, contained iron, rust, and occasionally had biological organisms of 2000 units. Viruses, because of their small size, would probably more easily become enmeshed in a protective coating of turbidity-contributing matter than bacteria would. For most effective disinfection, turbidities should be kept below 1 JTU; indeed, it would be best to keep the turbidity as low as 0.1 JTU, as recommended by AWWA water quality goals. The limit of 5 JTU of turbidity specified in the USPHS Drinking Water Standards, 1962, is meant to apply to protected watersheds and not to filtration plant effluents. With turbidities as low as 0.1 to 1, a preplant chlorine feed need be only enough to have a 1 mg/l free chlorine residual after 30 minutes contact time. Postchlorination practice would depend upon the ability to maintain such residuals throughout the distribution system.

In conclusion, there does not appear to be cause for panic or over-reaction to the problem of viruses in water. Under certain conditions, infectious hepatitis can be transmitted by treated water, but the evidence indicates that in such cases treatment was inadequate. The evidence for the transmission of other enteric viruses by treated water is, for the most part, speculative, with the possible exception of viral gastroenteritis and diarrhea, and nothing much is known about these viruses. These statements, however, do not mean we can be smug or complacent. There is still considerable room for research, both laboratory and epidemiologic to determine if there is a problem in virus disease transmission by water; to determine if the coliform index is always adequate (negative coliform test may not indicate freedom from viruses); to devise better techniques for measuring viruses in water; to develop a laboratory method of detecting small numbers of viruses in large volumes of water.

The preceding literature review points to several factors that may be incorporated in the design of water reclamation facilities to alleviate potential problems in water supplies containing large quantities of wastewaters.

An AWT plant (a good secondary sewage treatment plant followed by an efficient water purification plant followed by activated carbon adsorption) should remove virus based on the experience of numerous existing waterworks which receive raw water containing raw, primary, or secondary treated sewage. Actually the AWT plant will do more than this because the sewage treatment portions of the plant are designed and the plant may be operated with the specific aim of producing a safe water from wastewater. The addition of provisions for adsorption of organics on carbon is another plus factor.

In AWT plant design, provisions can be made so that a coagulant, alum, may be added continuously to the filter influent water following high lime treatment, in such a concentration that maximum filterability is assured. At times, a polymer may also be applied at this point. Mixed-media filters which will be even more efficient than the coal-sand media tested by Robeck, et al. in removing viruses, may be used. Several investigators have pointed out the close relationship between virus removal and turbidity reduction. The mixed-media filters at Tahoe have under test produced water with turbidity as low as 0.02 JTU. Ordinarily they are operated at about 0.2 JTU.

The excellent clarity of the water enhances chlorination efficiency and assures that viruses cannot escape chlorine contact by being encapsulated in particulate matter. Also the pH of the water should be lowered to about 7.0 prior to chlorination, to improve disinfection efficiency.

The high pH of 11.0 when lime treatment of secondary effluent is used also benefits virus removal in addition to its bactericidal effects. Recent tests by the Advanced Waste Treatment Research Laboratory of the EPA indicated that pH had little effect on virus at pH values of 10 or below. However, lime coagulation to raise the pH to 10.8–11.1 increased the rate of virus destruction rapidly. A pH of 11.1 was found sufficient to destroy polio virus in less than 1.5 hr.

In addition to all of the above treatment considerations which are common to the Tahoe project, AWT plant designs may also include breakpoint chlorination. The free chlorine residuals provided by this process are more effective disinfectants than the combined forms found at Tahoe. Thus even a higher degree of assurance of pathogen removal may be provided.

Between May 29 and October 2, 1969, 9 sets of Tahoe water samples were collected and submitted to the EPA laboratory in Cincinnati for virus examination under the direction of Dr. Gerald Berg. Each set consisted of 4 samples, 1 each from primary effluent, secondary effluent, carbon column effluent, and reclaimed water. Table 9-1 shows the results of the tests. In one sample of secondary effluent, there was

TABLE 9-1   VIRUS SAMPLING, 1969, AT THE SOUTH TAHOE PUBLIC UTILITY DISTRICT WATER RECLAMATION PLANT
Viruses Recovered (Plaque Forming Units)

	Primary Effluent	Secondary Effluent	Carbon Column Unchlorinated Effluent	Chlorinated Final Effluent
May 29	3	0	1	0
June 5			0	0
June 12			0	0
Aug. 20	3	18	NRD[a]	0
Aug. 27			NRD	0
Sept. 11			0	0
Sept. 18	179	14	9	0
Sept. 25	NRD	430	0	0
Oct. 2	207	320	0	0

[a]No reliable data.

no recovery of virus. In the other samples of secondary effluent, the counts varied from 14 to 430 units. Viruses were recovered from 2 of the 9 samples of carbon column effluent with counts of 1 and 9 units. After chlorination, all 9 tests of the reclaimed water were negative for virus. No virus has been recovered from the water being exported to Indian Creek Reservoir in 2 summers of sampling. While this does not necessarily mean that the water is entirely free of viruses, at least all of the results to date are favorable.

Complete virus removal of inactivation at Tahoe is finally dependent upon adequate chlorination. Again, this parallels bacterial removal which is also dependent on chlorination to complete the task. In both cases, complete disinfection cannot be successfully carried out without proper pretreatment of the water by clarification, removal of chlorine demanding substances, and adjustment of pH to a low range favorable for best use of chlorine. Successful disinfection in the presence of ammonia, the usual condition in treating wastewater, depends to a very great extent upon the use of rapid, violent, and thorough mixing of the chlorine solution and the water being treated, as well as adequate chlorine dosages. Less chlorine is required for viral and bacterial disinfection of wastewater than might be estimated based upon its ammonia content. At Tahoe, the application of only 2 mg/l of chlorine is adequate to produce the results reported in the presence of 2 to 15 mg/l of ammonia nitrogen. As discussed above, the use of breakpoint chlorination will provide an even greater assurance of virus removal.

Heavy Metals

Chapter 7 of this book is devoted to a discussion of removal of heavy metals from water supplies and wastewaters.

Trace Organics

Many organic compounds can be present in wastewaters including herbicides, pesticides, kerosene, benzene derivatives, and similar substances. Little is known about the health hazards of trace amounts of these materials in water. There may be a problem in heavy chlorination of secondary effluents containing these materials due to the formation of chloro-organic compounds which may be toxic to fish. However, through the use of AWT methods, there are at least two ways to reduce the contributions of wastewaters to this problem. One is to chlorinate the wastewater prior to granular activated carbon adsorption, in which case the chloro-organics are removed by the carbon. A second is to remove the organics by adsorption on the carbon prior to chlorination, in which case there are no organic concentrations of consequence to react with the chlorine. The success of the latter method has been well-demonstrated at Tahoe where a rainbow trout fishery has been maintained for 5 years in Indian Creek Reservoir. Since trout fingerlings are one of the most sensitive fish to the chloro-organic compounds, this is a favorable indication so far as aquatic life is concerned.

Also, as discussed in Chapter 8, the use of activated carbon in the water treatment plant may provide removal of trace organics in those cases where control of upstream sources is not adequate.

Biostimulants

With all of the public and professional concern which has been expressed in the press over the past few years about stimulation of algal growths and eutrophication of lakes due to the presence of excessive amounts of phosphorus and nitrogen in untreated or partially treated wastewaters discharged to the nation's waterways, there is no need to enumerate the ill effects or to otherwise belabor the problem here. As discussed previously, the AWT processes can reduce the nitrogen and phosphorus concentrations in wastewater discharges to the lowest levels achievable with available technology to at least minimize this one source of biostimulants.

COMBINING AWT WITH NATURAL TREATMENT PROCESSES

Many current discussions of water reuse compare the relative advantages and disadvantages of secondary treatment followed by natural purification versus advanced wastewater treatment only. Such comparisons are based on the illogical assumption that advanced wastewater treatment would not be preceded by secondary treatment nor followed by natural purification processes. This is an unwarranted and invalid conclusion. Any rational proposal for water reuse would include wherever possible, and certainly where needed, all available methods for water quality improvement; that is, secondary treatment, AWT and (not or) natural purification. No one could argue against this total combination of all available processes being better than any lesser combination. Taking advantage of the benefits of both natural and advanced processes is wise when there is any question about either being adequate alone.

Some of the natural purification processes include: prolonged storage, sedimentation, dilution, exposure to sunlight, reaeration, biological action, and temperature. Water is not a natural environment for intestinal bacteria or viruses and they die off with time. Storage alone can effect a 99 percent removal of coliform organisms. The benefits of dilution lie principally in the reductions of concentrations, if the receiving water is of better quality than water being discharged.

Reaeration and biological action may oxidize or remove substances, including 85 percent or more ammonia removal. For example, natural phenomena in Indian Creek Reservoir accounted for removal of an average of 86 percent of the ammonia present in the influent reclaimed water during calendar year 1969.

Natural processes, relative to plant treatment processes, are for the most part low-rate and slow-acting. They take place over a considerable time period and often over appreciable distances. Storage and separation in time and space between wastewater discharge and water reuse provides several advantages. They provide time for monitoring and testing of water quality before use and they afford mixing and some compensation for variations in plant efficiency. They also have tremendous aesthetic benefits which must not be overlooked.

Reclaimed water should be stored in a reservoir and released as far as practical upstream from the point of use in order to take advantage of natural purification and the other benefits.

The die-off rate of viruses in the aquatic environment is a complex phenomena which is dependent upon many factors. These rates vary with the type of virus as well as water temperature, nature of the water, and chemical characteristics of the water. Under otherwise identical conditions, virus survival is inversely related to temperature. Also, the die-off is directly related to the time the virus is in the water. Factors other than temperature and time have not been well-defined. Studies have been made on virus survival in sewage stabilization ponds. Polio virus Type 3 was found to be reduced by 99.99 percent in 96 hr in one case and by 99.9 percent in 72 hr in another. Still another study found a 99.5 percent reduction in polio virus Type 1 in 72 hr. Virus reductions of 99.9 percent were measured in detention times of 80–160 hr with polio virus Type 1 in another set of experiments. A review of the loss of infectivity of enteric viruses when suspended in various aquatic environments showed the time required for a 99.9 percent reduction at water temperatures of 15–16°C varied from 7–56 days, but at temperatures of 20–25°C, the time range was 2–28 days with most investigators reporting 99.9 percent reduction in several types of enteric virus in 8 days or less.

As discussed previously, several AWT treatment processes are particularly effective in removing viruses: coagulation with lime at high pH, filtration to provide very low turbidities, and breakpoint chlorination. However, should some viruses escape the AWT process, detention of the effluent in a clearwater impoundment will provide further assurance of virus removal prior to release of the reclaimed water. Likely significant reduction of nitrogen will also occur in effluent impoundments as observed in the Indian Creek Reservoir.

The use of the most effective available technology in both wastewater and water treatment facilities coupled with the separation by time and space of wastewater discharges with points of water use will enable satisfactory water quality to continue to be produced. In light of the increasing quantities and complexity of waste streams being discharged to water supplies, the time is rapidly approaching when any lesser approach may jeopardize the quality of treated waters in many areas of the country.

REFERENCES

1. Robeck, G. G., Clarke, N. A., and Dostal, K. A., "Effectiveness of Water Treatment Processes in Virus Removal, "*Journal American Water Works Association,* **54,** p. 1275 (1962).
2. Hudson, H. E., Jr., "High Quality Water Production and Viral Disease," *Journal American Water Works Association,* **54,** p. 1265 (1962).
3. Clarke, N. A., Berg, G., Liu, O. C., Metcalf, T., Sullivan, R. and Vlassoff, L., "Committee Report, Viruses in Water," *Journal American Water Works Association,* **61,** p. 491 (1969).
4. Clarke, N. A., Berg, G., Kabler, P. W., and Chang, S. L., "Human Enteric Viruses in Water: Source, Survival and Removability," *Advances in Water Pollution Research, Proc. Int. Conf., London, 1962,* Vol. 2, Pergamon Press, London, 1964.
5. Sanderson, W. W., and Kelly, S., "Discussion of Human Enteric Viruses in Water: Source, Survival and Removability," *ibid.*

10 Applications of New Technology to Existing and New Plants

INTRODUCTION

Many existing water treatment plants are faced with the concurrent problems of increased water demand, decreased raw water quality, and increased public concern over finished water quality. The treatment techniques described earlier in this book may, in many cases, be applied to existing plants to improve their performance while also increasing their capacity. By using these techniques, it may be possible to achieve these goals at costs substantially less than would be associated with construction of new facilities. These cost savings increase the probability of realizing needed improvements in the face of economic restraints. The purpose of this chapter is to describe full-scale applications of the techniques described earlier in this book to existing and new plants.

EXISTING PLANT EXPANSIONS

General Considerations

The expansion of existing plants by the installation of tube settlers or the conversion of rapid sand filters to mixed-media may enable the capacity of the treatment facility to be greatly increased without the need for any structural additions (i.e., new clarifiers or new filters). The majority of rapid sand plants are sound physically and are designed to lend themselves to the modifications necessary for such expansion. Modifications, however, may require ingenuity. As Culbreath has noted,[16]

> The companies that have developed some of the high-rate systems will guarantee rated capacities, that for most waters will approach three times the old-time rate of 2 gal/ft²/min for rapid sand filters. The top hydraulic rate of some of these plants can be increased as much as 300 percent and still produce higher quality water than that of the original rapid sand plant.
>
> The potential capacity, by use of high-rate filtration in the rapid sand plants in the U. S. is sufficient to last from one to two generations. This is significant.

Among the chief general considerations are the hydraulics of the existing rapid sand plant. The raw water pumping station will require expansion either by added construction or the replacement of existing pump(s) with larger units.

As discussed earlier, installation of tube settlers in an existing settling basin and subsequent increases in basin throughput will substantially increase the amount of sludge to be removed. Thus, the capacity of the sludge collection device and sludge pumps must be checked to insure adequate sludge removal capability.

The influent conduit is one hydraulic element that can rarely be changed in size since it is not practical to interrupt the plant operation for long periods to modify the conduit. One approach is to install a large auxiliary pipeline outside the plant parallel to the existing influent conduit and tie it in to the conduit at several locations near the major take-off points. The filter influent and effluent piping must usually be enlarged. The backwash supply and waste lines need not be altered since the backwash rates are not significantly altered by conversion to mixed-media. Filter rate controllers capable of handling the higher flow rates must be installed. Also, the plant clearwell and high-service pumps must be examined to be sure that they are compatible with the increased water demands prompting the plant expansion.

Throughout the remainder of this chapter, the term "mixed-media" is applied *only* to the tri-media filter discussed earlier.

Erie County, New York

The application of mixed-media to this plant in Erie County, N. Y. is of special interest because it provides side-by-side comparisons of the efficiency of mixed-media filters and rapid sand filters in a large plant. The results have been reported by Westerhoff[18] and are summarized below.

The Erie County Water Authority serves filtered water to approximately 390,000 persons in all or part of 12 towns and the city of Lackawanna in Erie County, N. Y. The average water production in 1969 was 44.3 mgd. Water is taken from Lake Erie, treated in the Water Authority's two filtration plants (Sturgeon Point and Woodlawn Filter Plants), and pumped into the transmission and distribution systems. A small portion of the total demand during peak periods is met by purchasing treated water from the city of Buffalo. Most of the water is treated at the Water Authority's Sturgeon Point Filter Plant, located in the town of Evans, approximately 30 miles south of the city of Buffalo. This plant originally was constructed in 1961.

In 1965, the water demanded by a burgeoning population in the suburban Buffalo area brought about expansion requirements. The first stage (to a capacity of 60 mgd) was needed immediately. A second expansion will be needed by about 1975, if the basis for forecasting remains valid.

The Water Authority decided during the first-stage expansion to install mixed-media in 6 new filters. The 4 original filters contain fine-to-coarse graded-sand media.

Special data were collected from a total of 48 parallel-filter runs during a 12 month study period. The sand filters were operated at the standard 2gpm/ft² filtration rate as a control. The filtration rate on the mixed-media filters was varied between 2–10 gpm/ft².

The water treatment processes at the Sturgeon Point Filter Plant consist of aeration, chemical addition, rapid mixing, and flocculation and sedimentation. These steps are followed by filtration and chlorination.

Flocculation and sedimentation are accomplished in 5 units. Each of the 5 units consists of a rectangular flocculation tank followed by a rectangular settling tank. Individual flocculation tanks have a liquid capacity of 46,500 ft³ (about 350,000 gal). At a capacity of 8 mgd, the detention time is nearly an hour. 2 of the tanks are equipped with mechanical paddle-wheel-type flocculators, and 3 of the tanks have walking-beam-type flocculators.

From the flocculators, flow passes through rectangular settling tanks, each having a surface area of 12,700 ft² and a volume of about 1.4 million gal. At a rated capacity of 8 mgd, the settling period is slightly longer than 4 hr with a surface loading of 630 gpd/ft².

The 4 original rapid-sand filters consist of a single-media bed of hydraulically graded fine-to-coarse sand. When rated at 2 gpm/ft², their combined capacity is 16 mgd.

The 6 new filters contain a mixed-media and have a total area of 8400 ft². The plant capacity, assuming a filtration rate of 2 gpm/ft² for all 10 filters, is 40 mgd. At a filtration rate of 3 gpm/ft², the plant capacity is 60 mgd.

During the parallel tests, the filter runs were terminated at a head loss of 8 ft. Typical turbidities applied to the filters were 2–4 JTU with little variation as the clarifier overflow rate was varied from 625–940 gpd/ft². Alum feed rate were typically 14–16 mg/l. The following conclusions were reached:

- More than 80 percent of the time, the filtered water turbidity from the sand filters was 0.10 JTU or less.
- Nearly 88 percent of the time, the filtered water turbidity from the mixed-media filters was 0.10 JTU or less.
- The mixed-media filters operating at filtration rates of 2–10 gpm/ft² consistently produced a lower filtered water turbidity than the sand media filters operating at a filtration rate of 2 gpm/ft².
- At filtration rates up to 6 gpm/ft², the mixed-media-filter effluent had a lower total microscopic count than the sand media filters operating at a 2 gpm/ft² rate.
- The mixed-media filters operating at 5 gpm/ft² had an average length of filter run of 29 hr. At 6 gpm/ft² they had an average filter run of 20 hr.
- The mixed-media filters operating at 5–6 gpm/ft² used a considerably lower proportion of wash water than the sand filters operating at 2 gpm/ft² (1.8 percent as compared to 2.5 percent on the average).

As a result of these tests, it was concluded that the mixed-media filters at 6 gpm/ft² produced an end product superior to that from the rapid sand beds at 2 gpm/ft² with concurrent economic advantage. The finished water goals adopted for filtered water at this plant were established for the mixed-media filters at 6 gpm/ft² as follows:

Turbidity	Less than 0.10 JTU average
	Less than 0.50 JTU maximum
Total microscopic count	Less than 200 su/ml average
	Less than 300 su/ml maximum
Color	Less than 1 unit
Odor	Not detectable
Aluminum	Less than 0.05 mg/l
Iron	Less than 0.05 mg/l

Greenville, Texas

In 1957, Greenville constructed a rapid sand filter plant with 4 filters with a nominal capacity of 6 mgd at a filter rate of 2 gpm/ft² (2167 ft² of total filter area). By 1965, water demands increased to the point that added capacity was needed. An evaluation of the economics

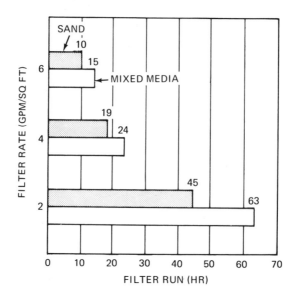

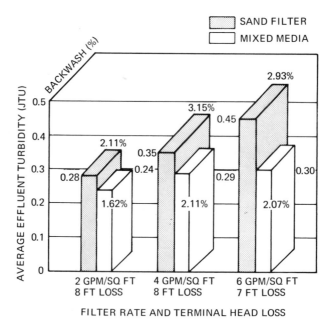

Fig. 10-1   Summary of results at Greenville, Texas: mixed-media *vs* rapid sand.[19]

of alternate expansion plans indicated that a conventional expansion to 10.5 mgd capacity would cost $475,000 while a conversion to mixed-media could be realized at a cost of $200,000.[12] The media in two of the sand filters was replaced with mixed-media, equipment to feed polymer to the filter influent was installed and a pilot filter coagulant control system was installed. As in Erie County, this plant modification provides side-by-side data for mixed-media filters at high rates as compared to rapid sand filters at conventional rates. Figure 10-1 summarizes the results of filters at different flow rates.

The data indicate mixed-media was superior to rapid sand filters when both were operated under the same conditions. This indicates that the mixed bed represented a fundamental improvement in this case. An alum demand curve was plotted for both filters, showing coagulant dosage against effluent turbidity under constant filter throughput. These curves indicated that less alum was needed for a given effluent clarity using mixed-media.

Both during normal plant operations and these special tests, it was reported[12] that operators came to place considerable confidence in the pilot filter as a control tool. It gave a good prediction of filterability within 5 min of the coagulation of incoming raw water. Considerable change in coagulant dosage could be made with the conviction that any subsequent adjustments required could be made long before trouble developed in the plant filter.

Even at the filter rate of 2 gpm/ft², the mixed-media showed impressive savings in wash water. At this rate, mixed-media would save this plant 26,000,000 gal of filtered water per year. At higher filter rates the savings would be greater.

Greenville has decided to convert the 2 remaining sand filters to mixed-media. Clarification capacity will remain unchanged, but improvements will be made in coagulation and flocculation. This means the nominal capacity of this plant will have been increased from 6 to 15 mgd and peak capacity will have been extended from 12 to 24 mgd without any structural additions to the plant.

Siler City, North Carolina

In the summer of 1962, it became apparent to the officials of Siler City, North Carolina, that it would be necessary to expand the capacity of their municipal water treatment plant in order to keep pace with the increasing demand for domestic and industrial water. The existing plant consisted of flocculation, sedimentation, and 4 conventional rapid sand filters with a total rated capacity of 1.00 mgd. This facility was overloaded and it was determined that the capacity should be doubled.

The old plant mixing basin was of the rectangular over-and-under baffle type, designed for 20 min detention based on a 1.0 mgd capacity. No change was made in the mixing basin except for removal of alternate baffles and the addition of a flash mixer at the influent end. Existing sedimentation basins were rectangular, designed for 4 hr detention. The 4 filter shells, each 9 × 10 ft were already equipped with Wheeler bottoms, cast iron wash water troughs and Palmer surface agitators. Except for the replacement of the filter media by mixed-media, only minor changes were made to the filters. Inlet and outlet valves were increased in size. New loss of head gauges and rate of flow instrumentation was installed and, in general, all piping, valves, and instrumentation throughout the plant were changed to accommodate a new design flow at 4.00 mgd.

The modified plant has been found[10] to produce a high clarity water at filtration rate of 2 1/2 times the previous plant rating. The rated capacity of the plant now is 2.50 mgd. Individual filter tests show that rates beyond the design are possible when future need develops. Successful tests have been run at rates up to 8.0 gpm/ft², and it is believed that the plant could be operated at 4.0 mgd and still economically produce an acceptable quality of water.

The cost of converting the existing 1.00 mgd plant to 2.50 mgd was approximately $40,000 less than the estimated cost of making a conventional 1.00 mgd addition. Actual cost of conversion was $138,000, whereas it had been estimated that addition of a 1.0 mgd conventional plant would have cost $178,000.

Siler City takes its raw water from the Rocky River. This stream is recognized to be a difficult water to treat in that it is constantly changing in character, with wide variations in turbidity, color, iron, manganese, and organic contamination. Typical operating results taken from the plant log at various times in the year are as follows:

Raw Water Turbidity Units	Filtered Water Turbidity JTU	Length of Filter Run hr.	Backwash Water Precent
140	0.1	43	1.5
350	0.1–0.3	35	1.8
125	0.1	42	1.5
120	0.1–0.2	69	1.5
125	0.1–0.2	24	2.5

Corvallis, Oregon

Corvallis has increased the capacity of its municipal water treatment plant by 2 1/2 times without increasing the size of the plant. Results of this application of advanced concepts have included a one third saving of construction costs, an improvement in water quality, and a considerable stretch-out of the city's long-range plans for expansion of its water supply.

Using the Willamette River as a source of supply, the original treatment plant was designed and built in 1949 as a conventional rapid sand filtration plant. It was designed in increments of 4 million gal per day capacity, each increment consisting of two 16 × 130 ft flocculation-sedimentation basins and 2 rapid sand filters, each with an area of 484 ft^2.

The initial plant capacity of 4 million gal daily rose to 8 mgd with the addition of a second increment in 1961. Under the original plan, the plant was scheduled for expansion to its ultimate capacity of 16 mgd with construction of the third and fourth increments in 1968–69. This, plus 4.5 mgd available from the city's second source on Rock Creek, was expected to take care of municipal water requirements through 1975, when construction of another treatment facility would have been necessary.

With application of new techniques to the existing plant, Corvallis has a water supply of 25 mgd, enough to last through 1980 at the current growth rate of 5 percent a year. Capacity of the Willamette River treatment plant has been increased without structural addition from 8 mgd to 21 mgd, and there remains space for the third and fourth increments called for under the original plan. Thus, the ultimate capacity potential has been boosted to 42 mgd. The expansion of capacity was achieved by application of (1) mixed-media filtration, (2) shallow-depth sedimentation with tube settlers, and (3) coagulation control techniques.

The turbidity of the raw water drawn from the Willamette River normally ranges from 15 to 30 JTU, with surges up to 1000 JTU. The water is soft (15–30 mg/l hardness) and exhibits periodic taste and odor problems. Typically, the following chemical dosages are used: alum, 20 to 40 ppm; lime, 10–20 ppm; chlorine, 2 ppm; polymer (as coagulant aid), 0.1–0.2 ppm; activated carbon, 5–10 ppm (for taste and odor control).

A diaphragm pump feeds alum into the existing flash-mix basin, 10 × 10 × 14 feet, in liquid form. The outlet of the flash-mix basin was modified to accommodate the increased hydraulic capacity.

Virtually all of the treatment piping had to be enlarged to handle the increased plant flow. From the flash-mix to the sedimentation basin, 2 welded steel pipelines were increased from 18 to 24 in.; from the sedimentation basin to the filters, one 24 in. line increased to 36 in.; in the filter gallery, the influent line increased from 16 to 24 in. and the effluent line, from 12 in. to 16 in. Butterfly valves have replaced all gate valves.

Settling tube modules were installed over about 60 percent of the 3500 ft^2 of rectangular settling basin area (see Figure 10-2). The tubes were installed on simple "I" beams spanning the width of the basins and are located at the discharge end of the basin. New effluent weirs and launders were also installed to provide good flow distribution through the area covered by the tubes.

Overflow rate in the sedimentation basin area covered by the tubes is 4.2 gpm per ft^2. This compares with the 1.05 gpm per ft^2 loading on the basin before they were installed.

Utilization of the existing basins with the tubes reduces the load on filters, especially during period of high river turbidity, and provides more economical operation by increasing filter runs during normal river conditions. With an extreme turbidity range of 2 to 1000 JTU, filter runs have been increased from 40 hr experienced with the old conventional basins at much lower plant throughput rates to 60–65 hr. Operating on raw water turbidities of 15 to 30 JTU, the tube effluent has had turbidities of 1 to 2 JTU.

The capacity of the filtration portion of the plant was increased by modifying the filter piping and by replacing the rapid sand filter media with mixed-media. 2 of the 4 filters had Leopold underdrains and the other 2 had Criscrete underdrains. At the 21 mgd rate, the filter rate is 7.5 gpm/ft^2.

The pilot filter coagulation control system described in Chapter 4 is used in conjuction with a Hach 1730 CR low-range continuous reading turbidimeter. A pilot filter effluent turbidity of 0.2 JTU is considered the maximum value associated with proper coagulation. The full-scale plant continuously produces a finished water turbidity of 0.2 JTU or less.

The cost to expand the plant from 8 mgd to 21 mgd capacity (including a new 5 mg reservoir, a new high-service pump station and a crosstown 16 in. transmission line to the new reservoir) was $430,000 (1969 prices). It was estimated that the same expansion through the addition of new settling basins and filters would have been at least $650,000. Thus, improved performance and increased capacity were achieved while realizing a substantial savings in cost.

Buffalo Pound Filtration Plant

The Buffalo Pound Filtration Plant is the primary source of water for the Saskatchewan, Canada cities of Regina and Moose Jaw. It is located near the southeast end of Buffalo Pound Lake, approximately 20 miles northeast of Moose Jaw and 37 miles west of Regina. The design capacity of the plant is 12 mgd (Imperial), approximately two thirds of which is allocated to Regina.

The plant consisted of 4 upflow clarifiers each 65 ft in diameter and 8 rapid sand filters. Raw water is drawn from Buffalo Pound, a shallow

Fig. 10-2 The modified clarifiers at the Corvallis, Oregon plant in operation. *(Courtesy Neptune Microfloc, Inc.)*

depth impoundment which supports extremely high algal counts (100,000 units/ml) during summer months.

In order to expand the capacity of the plant from 12 mgd to 30 mgd, tube settlers were installed over the surface of the upflow clarifiers, the

TABLE 10-1   COMPARISON OF TUBE CLARIFIER WITH CONVENTIONAL CLARIFIER PERFORMANCE, BUFFALO POUND, SASKATCHEWAN[4]

	Turbidity, JTU	
Time Sampled	Tube Clarifier	Conventional Clarifier
9:40	0.81	0.65
10:15	0.71	0.62
11:00	0.85	0.70
11:30	0.78	0.67
11:55	0.84	0.68
1:50	0.70	0.67
2:30	0.62	0.60
3:00	0.51	0.47
3:30	0.40	0.57
4:30	0.98	0.48

[4]Tube clarifier operating at 7.5 mgd
Conventional clarifier operating at 2.75 mgd
Chemical Dosages

Alum	115 mg/l
Lime	34 mg/l
Chlorine	21 mg/l
$KMnO_4$	1.0 mg/l

rapid sand filter media was replaced by mixed-media, and piping and valving changes were made. Original testing was conducted by modifying 1 filter and 1 clarifier. Figures 2-2 and 10-3 illustrate the tube settler installation technique adopted for final use.

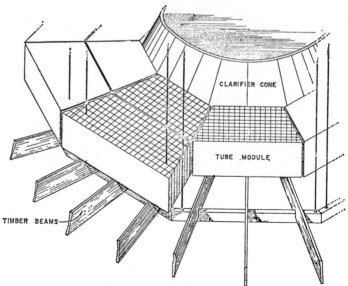

Fig. 10-3   Tube settler installation at Buffalo Pound (see also Fig. 2–2).[3]

Table 10-1 presents data obtained in one of these particular tests, where the modified clarifier was operated at a flow of 7.5 million gal/day in comparison with the conventional clarifiers at 2.75 million gal/day. The settling tube loading rate is 2 gpm/ft². The data illustrate that even though the tube clarifier was operating at over 2 1/2 times the rate of the conventional units, the effluent quality was comparable. Operating data were also obtained while a flow of 10.8 million gal/day (2.6 gal/min/ft²) was being passed through the converted clarifier. Even under these extremely high surface loading conditions, the clarifier continued to produce an excellent quality effluent which averaged 0.4 Jackson Units turbidity over a 3 hr test period.

Operating data accumulated to date reveal tube clarifier effluent turbidities ranging from 0.20 to 0.85 Jackson Units and averaging 0.6 Jackson Units at a flow of 7.5 million gal/day. For comparison, the effluent turbidity from the conventional clarifiers operating at 2.75 million gal/day have ranged from 0.30 to 2.20 Jackson Units, with an average of 0.70 Jackson Units over this same period.[7] Taste and odor indices for both units were found to be equal. The same high level performance is being routinely obtained with the converted plant filters at 5 gpm/ft², which are producing effluent turbidities ranging from 0.03 to 0.5 Jackson Units. Filter run lengths have ranged from 47 to 99 hr. The old sand filters frequently produced water containing turbidity in excess of 1 JTU and the backwash requirements approached 3 percent. Backwash requirements for the high rate system have averaged 0.7 percent.

An accumulation of flocculated material on the upper surface of the tube modules was noted. The floc continued to accumulate until it bridged over and completely covered the surface of the modules. The clarifier effluent quality was not affected by this accumulation. However, it was feared that the taste and odor of the water might deteriorate if the floc were not removed. The floc buildup appeared to be related to the quantity of algae in the raw water as the buildup ceased in the winter months, only to occur again in the summer months. Cleaning is now accomplished with a surface jet (similar to that shown in Figure 2-12) consisting of a spray bar radiating out from the clarifier center to the wall at 90° intervals. Biweekly operation of the surface jet prevents a significant floc buildup. The consulting engineer (Associated Engineering Services of Regina, Saskatchewan) reports successful use of tube settlers in 4 other plants and mixed-media in 12 other plants.

Newport, Oregon

A conversion similar to that made at the Buffalo Pound plant was completed in late 1967 at the Newport, Oregon, water treatment plant.[5] The existing 1.5 mgd plant consisted of a circular flocculator-clarifier followed by rapid sand filters. The plant was expanded to 3 mgd capacity by installation of tube settling modules in the clarifier and replacement of the rapid sand media with mixed-media.

The clarifier conversion was made by installing a ring of tube modules extending inward approximately 60 in. from the peripheral overflow weir and separated from the uncovered area by a baffle intersecting the surface of the clarifier. Approximately one third of the clarification area was covered with tube settlers. The resulting flow rate through the area covered by tubes was 5 gpm/ft². If additional capacity is required in the future, more of the surface area can be covered with tube modules.

Prior to the introduction of tube modules, filter runs through the existing rapid sand filters averaged 25 to 35 hr. Installation of tube modules resulted in a twofold increase in the average length of filter runs to 65 to 70 hr. Shortly after the clarifier modification was completed, the existing media was replaced with mixed-media, and a new raw water pump was installed to boost the plant flow from 1.5 to 3 million gal/day. This 100 percent expansion of the treatment capacity was attained without construction of additional plant structures.

A review of plant operating data for the months of August and September, 1968, reveals that the converted tube clarifier and the mixed-media filters were performing exceptionally well at a flow of 2.5 to 3 million gal/day. Raw water conditions during these months generally are the most severe of the year and impose the most difficult treatment problems. Table 10-2 is a summary of plant operating data taken during this period, which illustrates the performance of the plant at double the design rate. Data taken at a full plant capacity of 3 million gal/day showed the tube clarifier to be removing an average of 87 to 91 percent of the flocculator solids measured as turbidity. Filter run length, which has been averaging 17 hr, was expected to increase after improvements to the plant filter hydraulics and after flow controllers were completed, allowing operation of the filters to higher head losses. Optimization of the use of polyelectrolyte as a filter aid should also improve the filter efficiency by allowing full utilization of the floc storage capacity of the filter bed. Filter runs approaching 20 to 25 hr should be possible after these measures are completed.

Crossett, Arkansas

Another example of increasing the capacity of water treatment plant by utilizing tube clarifiers and mixed-media is Georgia-Pacific's Pulp and Paper Mill at Crossett, Arkansas.[6] This plant was expanded without construction of additional structures to provide the

TABLE 10-2 SUMMARY OF MONTHLY OPERATING DATA, AUGUST TO SEPTEMBER; 1968, NEWPORT, OREGON, WATER TREATMENT PLANT[a,b]

Plant flow	2.5–2.0 mgd
Raw water turbidity	10–15 JTU
Raw water color	160–200 units
Tube clarifier turbidity	1.5–2.0 JTU
Filtered turbidity	0.01–0.1 JTU
Chemical feed	
Alum	38 mg/l
Lime	14 mg/l
Polyelectrolyte	0.55 mg/l
Carbon	2–5 mg/l

[a]Average length of filter runs—17 hr to 6 ft of water head loss.
[b]Percentage backwash water—2.7 percent.

desired 45 mgd capacity. The original plant capacity was 30 mgd. The 6 plant filters received clarified water from two 120 ft diameter Infilco accelators which were upgraded from their original design capacity of 15 mgd each to 22.5 mgd each by installing tube modules (2 ft long tubes) over a portion of the clarifier surface. At the expanded flow of 45 mgd, the rise rate through the tubes is approximately 3.2 gpm per ft^2 of tube entrance area. This compares to the original design overflow rate of less than 1 gpm per ft^2 for the conventional basins. Tube settling pilot tests conducted during the winter of 1968 revealed that tube overflow rates of 4 gpm per ft^2 would provide effective clarification even under cold water conditions.

To take advantage of the flow distribution provided by the radial launders, the tube modules were installed by suspending them from these launders as illustrated in Figure 2-10. The 10,000 ft^2 of tube modules were installed in less than one week without removing the clarifiers from service.

Qualifying tests were completed during which the flow through one clarifier was steadily increased from the original operating rate of 12 mgd to a peak of 23 mgd. At the expanded flow, the clarification efficiency was superior to that obtained prior to the tube installation. Turbidities applied to the converted mixed-media filters were less than 1 JTU, resulting in filter runs in excess of 48 hr. Prior to installation of the tube modules and mixed-media filtration materials, filter runs at flows of 12 mgd to 15 mgd per clarifier ranged from 12 to 18 hr. In addition to the savings to the paper-making operation resulting from expanding the clarifiers without removing them from service, substantial savings were achieved in the plant expansion. The added 15 mgd to clarifier and filter capacity was achieved for less than $15.000 per mgd.

Fairfax County, Virginia

The Fairfax County, (Virginia) Water Authority supplies water to about 500,000 people in the suburban Washington, D. C. area. The area served is one of the most rapidly growing areas in the United States and the time required for construction of added, conventional treatment facilities was not compatible with meeting the increasing water demands.

The raw water supply is the Occoquan Reservoir which has an impounded watershed of approximately 570 square miles. The reservoir of 9.8 billion gal has a dependable yield of 65 mgd. Upstream discharges of secondary sewage effluent from several sewage treatment plants have contributed to the eutrophication of the reservoir and taste and odor problems become severe during summer months.

Water treatment is provided at two interconnected plants. The treatment units consist of 17 circular steel combination coagulating, settling and filtering units which had a nominal capacity of 46.4 mgd based on a filter rate of 2.0 gpm/ft². The maximum rated capacity based on 1.5 times the nominal rate was 69.6 mgd. Principal chemicals used in the treatment process are liquid alum, pre-and post lime, pre-and post chlorine, fluoride, activated carbon, and potassium permanganate.

The existing units at the River Station Treatment Plant are circular steel units exactly 86 ft. in diameter. The center portion is an upflow mixing, flocculation, and settling tank. The filter unit extends around the outer circumference. Original design features of the unit furnished by Dorr-Oliver, which was modified to test performance at higher filter rates, are as follows:

Nominal rating = 4.0 mgd at 2.0 gpm/ft²
Filter area = 1400 ft²
Settling tank area = 3848 ft²
S.W.D. = 17 ft 3 in.
Overflow rate = 1040 gpd/ft²
Detention time = 2.98 hr at 4.0 mgd
Wash rate = 15 gal/ft²/min

In September, 1968, the authority entered into a contract with Neptune Microfloc, Inc., for furnishing and installing tube settler modules in Unit No. 1 of the River Station Treatment Plant. In addition, the existing media, consisting of sand and anthrafilt, was to be removed and replaced with an inverse graded bed of ilmenite, sand, and coal. The contract documents required certain performance prior to acceptance by the authority as follows:

- Unit No. 1 would have to produce 10.0 mgd continuous flow with effluent characteristics at least equal to the averages for the other 4 units at 4 mgd during the period January 1, 1968 through August 31, 1968.
- Maximum rate of backwashing shall not exceed 15.0 gpm/ft² of filter surface area.
- Filtered water turbidity shall not exceed 0.2 JTU at 5.0 gpm/ft². Head loss shall be 8.0 ft or less.
- Backwash water shall not exceed 3 percent when water applied to the filter has a turbidity not in excess of 10.0 JTU.
- At the 10.0 mgd rate (5.0 gpm/ft²) the turbidity, odor and taste of water leaving the tube modules of Unit No. 1 shall be less than or equal to the turbidity, odor, and taste of water leaving Units 2, 3, 4 and 5 when they are operating at 4.0 mgd (2.0 gpm/ft²). The applied turbidity to the filter shall never exceed 10.0 JTU.

In addition, the Virginia Department of Health placed the following requirements on the project:

- The addition of chemicals and coagulation of the raw water would be controlled by zeta potential and a zetameter.
- Continuous indicating and recording turbidimeters would be used to monitor and record raw water, applied water, and filter effluent turbidity.
- Applied water should have an average turbidity of 2.0 JTU and a maximum of 5.0 JTU. Filtered water would have an average turbidity of 0.1 to 0.2 JTU with a maximum of 0.5 JTU.
- The unit would be operated continuously for a full year in parallel with the other 4 existing units in order to compare performance results.

In order to increase the capacity of the settling portion of the treatment units (Dorr-Hydro-Treators), tube settling modules were installed around the outer periphery of the settling tanks. The filter media in the test unit (Unit No. 1) was replaced with 18 in. of coal (1.0 mm effective size), 7 in. of sand (0.5 mm effective size), and 3 in. of ilmenite (0.18 mm effective size). The 4 standard units contained dual media filters made up of 24 in. of coal (effective size 1.0 mm) and 6 in. of sand (effective size 0.5 mm). Hach CR surface scatter turbidimeters and Hach CR turbidimeters were installed in the pipe gallery below the Hydro-Treators to record raw water, applied, and effluent turbidities. Some difficulties were encountered in the original testing due to the critical nature of the tube settler location in relation to the sludge blanket maintained in the units. Although no added published data were yet available at the time of this writing, it was reported that testing subsequent to the preliminary report[8] proved satisfactory and additional plant clarifier-filter units were modified in a manner similar to that described above.

Winnetka, Illinois

At the time of this writing, over 150 existing plants had been modified by the installation of mixed-media alone. A typical example is offered by the modifications of the Winnetka, Illinois plant. The existing 6.0 mgd plant was in need of expansion to 12.0 mgd but space limitations made such an expansion by conventional means very difficult. 4 of the 8 plant filters were originally converted to mixed-media, providing an added 3 mgd of capacity at a cost of $86,692.36 (see Table 10-3) in 1967. The estimated cost of providing an added 6

TABLE 10-3   COST TO CONVERT FOUR FILTERS TO MIXED-MEDIA, EACH 260 SQUARE FEET IN AREA (1967 PRICES)

Neptune Microfloc Corporation		
Filter material	$18,910.00	
Freight	2,103.19	
Control center	19,300.00	
Secondary flocculent equipment	4,480.00	
Secondary flocculent assembly TB-1	4,480.00	
Freight	527.57	
Technical assistance	800.00	
Miscellaneous expense	203.37	
Total Microfloc	$46,784.14	
BIF Division of New York Air Brake Company		
Four rate of flow tubes with butterfly		
valves and miscellaneous equipment	$ 8,571.00	
Allis Chalmers 30 in. valve	1,324.00	
Gallaher and Speck to install 30 in. valve	1,657.79	
Miscellaneous material	7,845.04	
Total other equipment and material	$19,397.83	
Total equipment and material		$66,181.97
Labor and equipment rental		
Village labor—remove and install		
filter media	$11,835.68	
Village labor—piping changes	2,633.77	
Village labor—electrical work	3,673.83	
Village labor—treatment equipment changes	1,764.03	
Total village labor	$19,907.31	
Consultant fees	$ 414.98	
Conveyor rental	188.10	
Total labor and equipment rental		$20,510.39
Total cost of conversion		$86,692.36

mgd of capacity by construction of added rapid sand filtration capacity was $950,000. Table 10-4 presents a summary of operating data collected over an 8 month period. Raw water is drawn from Lake Michigan. In addition to the media conversion, a pilot filter coagulation control center (see Chapter 4) was installed. The alum dosages previously used were reported to be reduced by 40 percent as a result of the more positive control offered by the control center — providing a savings of $1.50/mg.

APPLYING TECHNOLOGY TO NEW PLANTS

The fact that coarse-to-fine, in-depth filters as exemplified by mixed-media beds can store such large quantities of particulates which have been removed from the water without excessive head loss or floc breakthrough open up possibilities of several plant flow sheets other than the standard approach of full rapid mixing, flocculation, and settling as treatment prior to filtration. One of these is direct filtration of low turbidity coagulated waters without the need for sedimentation. This has been done in several applications. In general, direct filtration can be applied when the average raw water turbidity is in the range of 25 to 50 JTU.

Occasional turbidity peaks of 100 JTU can also be handled without difficulty. Another flow sheet for low turbidity waters provides a contact basin with 30 to 60 min holding time but without installation of equipment for the continuous collection and removal of sludge. The addition of the contact basin increases the potential range of turbidities which can be processed to about 20 to 100 JTU on the average with tolerable peaks as high as 200 to 1000 JTU.

Richland, Washington

The Richland, Washington plant is an all new city-owned plant deriving its supply from the Columbia River. Raw water turbidities average about 7 JTU (usual range 3 to 150 JTU) with peaks up to 1500 JTU. Water temperatures vary from 2° to 20°C. Alkalinity is about 50 to 80 mg/l, and coliform mpn 3.6 to 4600 per 100 ml in the raw water. Plankton are present in late summer and fall.

Briefly, the treatment process consists of the addition of chemicals (alum and lime), hydraulic rapid mixing, contact (1¼ hr), addition of a polymer (0.01 to 0.10 mg/l), filtration through mixed-media beds with Wheeler underdrains at 5 gpm/sf, and chlorination. The coagulant control center is used to adjust alum feed as necessary to maintain finished water turbidity at or below 0.20 JTU. This plant has been supplying potable water to the City of Richland since its completion in 1963. An exterior view of this plant is shown in Figure 10-4.

TABLE 10-4   COMPARISON OF RAPID SAND VERSUS MIXED-MEDIA FILTRATION MARCH TO NOVEMBER 1967

Filter Number	Filter Media	Filter Rate in Gallons Per Ft_2 Per Min	Hours of Operation	Number of Back Washes	Filter Run Average	Gallons Filtered	Gallons Back Wash	Percent Back Wash
4	Sand & Gravel	2.39	319 hrs 40 min	20	16 hrs 0 min	11,966,900	354,660	3
7	Mixed Media	2.38	322 hrs 15 min	9	35 hrs 48 min	12,020,300	170,000	1.4
8	Mixed Media	5.26	316 hrs 55 min	21	15 hrs 5 min	26,070,700	415,850	1.6

This summary is based on five different test runs over a period of eight months.
The water conditions and chemical feed ranges were as follows:
Water temperature 34°–65°C   Water turbidity   5–70   Loss of head 7.0 ft
Alum feed   47–125.1 lb per million gal   Pre-chlorination 11.8–15.5 lb per million gal
Carbon feed   19.3–31.9 lb per million gal   Post-chlorination 0.96–2.2 lb per million gal

Fig. 10-4   Exterior view of the Richland, Washington water treatment plant.

A detailed description of this plant is presented to illustrate several considerations involved in new plant design. Since many of these considerations are common to subsequent examples, they are not repeated for other plants subsequently discussed.

Raw water is supplied to the plant from an intake on the Columbia River. There are 2 vertical turbine pumps of 7.5 and 15 mgd capacity. The pumps may be started or stopped by push buttons operated locally at the intake or remotely from the plant control panel. The 36 in. discharge line from the intake to the treatment plant is designed for a future capacity of 45 mgd.

The raw water influent line discharges into a Parshall flume located above operating floor level in the treatment plant. The flume serves two purposes: to measure flow and to mix hydraulically the raw water and treatment chemicals. The raw water flow is indicated, recorded, and totalized on an instrument in the plant control panel. Water from the flume is discharged through a single conduit designed to carry up to 45 mgd to the settling basins.

Chlorine is delivered to the plant in ton containers. The chlorine storage and feed rooms have only outside entrances. Air is exhausted from these rooms by means of the unit heater fans which operate continuously at all seasons of the year. Storage space is provided in the plant at ground level for 7 containers. A 2 ton electric cable hoist and monorail handles the containers. The chlorine scale accommodates 2 cylinders.

There are 2 chlorinators, Wallace & Tiernan, Series A711, V-notch, each with a capacity of 25 to 500 lb per day, and each capable of feeding up to 2000 lb per day with installation of larger rotameters and orifices. Each is equipped for remote manual control. Three points of chlorine application are provided: to the raw water, to the separation bed influent, or to the separation bed effluent.

Liquid alum is added to the raw water in the 36 in. riser pipe to the Parshall flume. In addition, alum may be applied to the separation bed influent. Presently, the liquid alum is stored in a single rubber-lined, 6400 gallon steel tank, located on the ground floor level of the plant. Space is provided for the addition of a second alum storage tank of the same size.

Four alum feed and metering pumps are installed, 2 small and 2 large capacity units. Variable speed drive and variable stroke adjustments provide a range of alum feed of 1/4 to 123 gal per hr. Dilution water is added to the alum feed pump discharge to minimize line plugging problems, to reduce delivery time to point of application, and to aid mixing with the water being treated.

A coagulant control center is used to test the filtrability of the water. The turbidity of the control bed effluent is continuously measured and recorded to hundredths of a unit. Turbidity of 0.20 is taken as the control value. If the control bed turbidity exceeds 0.25, the alum feed is increased, and if it is less than 0.15, it is decreased. Under most operating conditions, all that is required is to adjust the alum feed to hold the turbidity within the control range of 0.15 to 0.25 unit.

At flows of 15 mgd, 1¼ hr detention is provided in the settling basins. The principal function of these basins is to effect some removal of suspended materials when the raw water turbidity is high. They also provide holdup or lead time for the control bed, although 10 min would serve this purpose. Operating experience to date indicates that the basins are more than ample in size for use with this process, and it is doubtful that additional basins will be built when the plant is expanded in the future.

The basins are equipped with an ice protection system which consists of a blower located on the ground floor of the building and perforated pipes installed around the periphery of the basins below the water level. Air bubbling from the pipe openings to the water surface prevent formation of an ice layer in the basins.

From 0.01 to 0.10 ppm of a polyelectrolyte is added to the separation bed influent water. Dosage requirements usually fluctuate slowly; the amount needed increases with higher flow rates through the beds, lower water temperatures, higher alum dosages, and higher turbidity in separation bed influent water. 2 solution feed and metering pumps are installed for polymer feed. Each pump has a maximum capacity of 38 gal/hr, and each is equipped with variable speed drive and variable stroke adjustment. The stroke settings are made at the feed machines, while the speed settings are made at the console in the control room.

Lime may be added to the water for any or all of three purposes: to adjust the calcium carbonate stability of the water to minimize corrosion; to assist coagulation by raising the pH to a favorable range; or to assist coagulation by supplying alkalinity at times of low natural alkalinity in the raw water or at times when high alum dosages are necessary.

Lime is applied to the raw water in the Parshall flume when coagulation assistance is needed, and to the separation bed influent when only corrosion control is required. Storage space is provided for about 45 tons of hydrated lime which is purchased and stored in bags. The lime hopper is equipped with a dust collector, and an incinerator is provided near the hopper for disposal of paper bags.

Presently, a single volumetric dry lime feeder with a capacity of 20 to 400 lb/hr of dry lime is installed, but space is available for additions. Powered activated carbon is purchased in bags and handled in the same manner as lime. Storage space is provided for 22 tons of carbon. A single dry feeder with a range of 4 to 80 lb/hr is installed, but space is provided for more units when needed. Carbon may be applied either to the raw water in the Parshall flume or to the separation bed influent.

There are 4 mixed-media filters each having a nominal capacity of 3.75 mgd at 5 gpm/ft². With the presently installed rate-of-flow controllers, the maximum rate of operation possible is 5 mgd per bed (6.7 gpm/ft²) or a total of 20 mgd through the plant. Actually, the plant is operated quite often at these higher rates, and effluent quality is maintained. All parts of the beds and the piping are designed hydraulically for flows up to 7.5 mgd per bed (10 gpm/ft²), or 30 mgd through the 4 units. For flows above 20 mgd, however, it will be necessary to increase the size of the rate controllers.

The filters are equipped with Wheeler bottoms and Palmer rotary surface washers. Backwash water is supplied by a vertical turbine pump having a maximum capacity of 10,500 gpm. The beds may be operated up to 10 ft of head loss before backwashing. Filter head loss is indicated on the control console, and the turbidity of the bed effluent is recorded at the coagulant control center. The beds are provided with an ice protection system similar to that in the contact basins. The filters are not housed, nor is the areas above the pipe gallery between the opposed pairs of separation beds, since all of the operating controls are located in the laboratory-control room.

An existing 2.2 mg reservoir at the plant site was converted to a clearwell. It provides equalization between plant output and high service pumpage and storage of wash water.

There are 2 new vertical turbine pumps with capacities of 2700 and 5400 gpm, and 2 old horizontal centrifugal units, each with a capacity of 5000 gpm. The firm capacity with the largest pump out of service is 18.3 mgd. Pumpage to the distribution system is measured by a Sparling meter and recorded on a chart in the control room

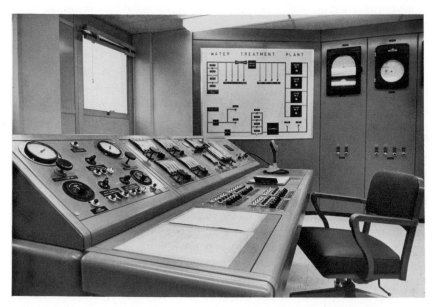

Fig. 10-5   Filter control console at the Richland, Washington water treatment plant.

This is a plant which is truly controlled and operated from the laboratory. The laboratory and control equipment are located in the same room. The console contains all of the controls for operation of the separation beds and adjustment of all chemical dosages. A graphic panel gives the flow pattern and indicates by lamps which plant units are in operation (see Figure 10-5). The plant control panel also contains a coagulant control center; a pH recorder; raw and treated water flow recorders; a clearwell depth indicator; low and high service pump controllers; and the plant discharge-main pressure recorder. The laboratory-control room also contains a panel for supervisory control of all outlying parts of the pure water system including reservoirs, pump stations, and wells.

Use of the coagulant control center has been described previously, and some of the laboratory tests for plant control and record purposes have been mentioned under various sections relating to plant process units. In summary, automatic equipment is used for testing and recording pH of the raw, treated, and stabilized water, and turbidity of plant and pilot bed effluent, and for periodic tests of chlorine residual, qualitative taste and odor, alkalinity, raw water turbidity, and temperature. In addition, bacteriological samples are collected regularly and submitted to the health department for analysis.

The plant at Richland has been in operation since June, 1963, under the direction of John A. McCool, Water Superintendent, and Russell A. Stephens, Chief Operator, and the results to date have been excellent.

Medford, Oregon

The Medford, Oregon facility is an all new municipally-owned plant with a capacity of 15 mgd. It takes its supply from the Rogue River. This is a pumped supply which is used in the summer months as a peaking plant to supplement the base supply which is obtained by gravity flow from mountain springs. The treatment of Rogue River water consists of rapid mixing of alum by a Parshall Flume, 1 1/4 hr retention in a contact basin, addition of a polymer, mixed-media filtration at 5 gpm/sf (Wheeler underdrains), chlorination, and dechlorination with SO_2. The plant has served the city of Medford with high quality water (turbidity 0.1 to 0.2 JTU) during the summer months since 1968. An exterior view of this plant is shown in Figure 10-6.

Fig. 10-6   Exterior view of the Medford, Oregon water treatment plant.

Arnprior, Ontario, Canada

The Arnprior, Ontario 4 mgd plant takes its water supply for municipal purposes from a surface source which has a color of 20 to 70 and a turbidity of 2 to 100 JTU. This plant was placed into service in 1968. The treatment consists of addition of 20 to 30 mg/l of alum and 5 to 15 mg/l of soda ash, rapid mixing, 1 hr contact time, mixed-media filtration at a normal rate of 3.4 gpm/sf and a maximum rate of 7 gpm/sf, and chlorination. The finished water has a color of 0 to 5 units and a turbidity of less than 0.2 JTU.

Peoria, Illinois

The all new plant at Peoria, Illinois is owned and operated by a private company. Since 1965, it has provided up to 10 mgd of water from wells for municipal use following treatment in an iron and manganese removal plant. The well water supply contains from 0.1 to 2.5 mg/l of iron, and 0.3 to 1.0 mg/l of manganese. About 2.5 mg/l of chlorine and 1.0 mg/l of potassium permanganate are added to the raw water and 30 min of contact are provided. About 0.013 mg/l of a polymer is added to the filter influent. There are 5 pressure filters which operate at 5 to 13 gpm/sf. The filters are mixed-media with pipe lateral underdrains. The concentrations of iron and manganese in the finished water are each less than 0.05 mg/l.

Filter backwash is fully automatic and is also automatically initiated based on high head loss, high turbidity, excess iron, or by time clock. Backwash water use varies from 1.5 percent at 5 gpm/sf to 2.75 percent at 8 gpm/sf, and the length of filter runs varies from 8 to 20 hr.

Clackamas, Oregon

The Clackamas, Oregon facility is an all new municipally owned plant which supplies up to 8 mgd for public use. The water source is in the Clackamas River which has average turbidities of 20 to 50 JTU with surge peaks of from 500 to 1000 JTU. The treatment process includes 1 min flash-mixing, a 1 hr contact basin, and gravity mixed-media filters with Wheeler underdrains. The filters normally operate at 6 gpm/sf. Typical chemical dosages are: alum 10 to 20 mg/l; lime 3 to 7 mg/l; polymer 0.05 mg/l; and chlorine 1 to 2 mg/l. Filter runs are from 20 to 30 hr. Finished water turbidities are held to 0.1 to 0.2 JTU by means of a coagulant control center. In summer, at times of low raw water turbidities, filter rates have been as high as 8 gpm/sf.

Lake Oswego, Oregon

The Lake Oswego, Oregon 11 mgd plant takes its supply for municipal use from the Clackamas River which has turbidities nor-

Fig. 10-7   Exterior view of the Lake Oswego, Oregon, water treatment plant

mally in the range of 20 to 50 JTU with peaks up to 1000 JTU. The treatment process consists of addition of alum at 20 to 30 mg/l and lime 5 to 10 mg/l, 1¼ hr contact, polymer at 0 to 0.05 mg/l, filtration at 5 gpm/sf through mixed-media with Leopold clay tile filter underdrains, and chlorination. Finished water turbidities are 0.1 to 0.2 JTU. Filter runs vary from 20 to 30 hr in winter to 50 to 70 hr in summer. Figure 10-7 is an exterior view of this treatment plant.

Aurora, Colorado

The Aurora, Colorado 20 mgd municipal plant takes its supply by way of a transmountain diversion from the upper Eagle River. Raw water turbidities vary in summer from 10 to 40 JTU and in winter from 1.5 to 6.2 JTU. Treatment consists of addition of about 4 mg/l of alum in winter and 18 mg/l in summer, hydraulic rapid mix, 1 hr retention in a contact basin, addition of about 6 mg/l of caustic for pH control, polymer as a filter aid at 0.03 mg/l, addition of about 0.7 mg/l of powdered carbon when required for taste and odor control, mixed-media filtration at 5 gpm/sf (hydraulics for 9 gpm/sf) with Wheeler underdrains, and chlorination. Finished water turbidities range from 0.2 to 0.5 JTU. This plant is planned for expansion up to 80 mgd capacity.

This plant adds powdered activated carbon at times to the filter influent for taste and odor control. This points up an important ability of mixed-media filters to retain powdered carbon without leakage. In other plants, carbon dosages up to 50 mg/l have been added ahead of mixed-media filters without the appearance of carbon in the effluent. The retention of carbon particles in the filter bed has little effect in increasing head loss even at high dosages.

Arvada, Colorado

The Arvada, Colorado 18 mgd municipal plant purchases its supply from Denver's Moffat system which is a transmountain diversion from the West Slope of the Rockies. Raw water turbidity is normally 10 to 25 JTU with high organic color in summer and with occasional turbidities as high as 800 JTU. Treatment consists of addition of 10 to 16 mg/l of alum, 6 to 8 mg/l of lime, hydraulic rapid mix, 36 min contact, addition of 0.08 to 0.2 mg/l of polymer as a filter aid, filtration through mixed-media at 5 gpm/sf (Wheeler underdrains), and chlorination. Finished water turbidities are maintained at 0.2 to 0.4 JTU. Backwash water use amounts to 3 percent of production.

Las Vegas, Nevada

The Southern Nevada Water System includes a 200 mgd direct filtration plant which draws its supply from Lake Mead. The plant includes 10 dual media filters (20 in. of anthracite and 10 in. of sand) designed for a 5 gpm/ft^2 maximum filter rate. The plant was placed in operation in 1971 and was constructed at a cost of $5,800,000 (1969 prices). In addition to coagulants, powdered carbon is fed ahead of the filters for taste and odor control. At the flows of 100 mgd being experienced in 1973, filter runs of 80 hr were achieved at raw water turbidities of 1–3 JTU.[20]

Springfield, Massachusetts (Proposed)

In Springfield, Massachusetts, there are plans to use dual media filters in a 60 mgd plant in much the same manner as in the plants described just previously which employed mixed-media. Pilot tests were used to develop the following basis of design: the addition of lime and alum, rapid mix, 25 min of mechanical flocculation, polymer addition if needed, dual media filtration at 5 gpm/sf (24 in. of 1.0 to 1.1 mm coal on 12 in. of 0.45 mm silica sand) and Camp-nozzle tile block underdrains, caustic addition, and chlorination. The new system will be operated in parallel on the same supply with existing slow sand filters which have been in service for more than 70 years. Bids on the project indicate that this type of direct filtration plant can be built at less than half the cost of a conventional plant using rapid sand filters preceded by settling basins.

REFERENCES

1. Collins, F., and Shieh, C. Y., "More Water from the Same Plant," *The American City*, p. 96 (October 1971).
2. Eunpu, F. F., "High Rate Filtration in Fairfax County, Virginia," *Journal American Water Works Association*, **62**, p. 340 (June 1970).
3. Livingston, A. P., "High Rate Treatment Evaluated at Buffalo Pound Filtration Plant," *Water and Sewage Works* (June 1970).
4. Hansen, S. P., Richardson, G. N., Hsiung, A., "Some Recent Advances In Water Treatment Technology," *Chemical Engineering Progress Symposium Series*, **65**, p. 207 (1969).
5. Culp, G. L., Hansen, S. P., and Richardson, G. H., "High Rate Sedimentation in Water Treatment Works," *Journal American Water Works Association*, **60**, p. 681 (1968).
6. Culp, G. L., Hsiung, A., and Conley, W. R., "Tube Clarification Process, Operating Experiences," *Sanitary Engineering Division Journal*, American Society of Civil Engineers, SA5, p. 829 (October 1969).
7. Sedore, J. K., "We Converted to High Rate Filtration," *American City* (April 1968).
8. Westerhoff, G. P., "Experience with Higher Filtration Rates," *Journal American Water Works Association*, p. 376 (June, 1971).
9. Price, B., "First Major Water Treatment Advance in 67 Years," *American City*, (February, 1962).
10. Alley, L. A., "Water Plant Capacity Doubled Without Additional Construction," *Public Works*, p. 106 (Feb., 1965).
11. Culbreath, M. C., "Experience with Multimedia Filter," *Journal American Water Works Association*, p. 1014 (Aug., 1967).
12. Laughlin, J. G., and Duvall, T. E., "Simultaneous Plant-Scale Tests of Mixed-Media and Rapid Sand Filters," *Journal American Water Works Association*, p. 1015, (Sept., 1968).
13. Livingston, A. P., "High-rate Clarification and Filtration Augment Buffalo Pond Plant Capacity," *Water and Sewage Works* p. 119 (April, 1969).
14. Westerhoff, G. P., "Expansion of Suburban Buffalo Filtration Plant," *Water and Wastes Engineering*, p. 68 (June, 1970).
15. Harris, G. A., and Ringwood, R. J., "Water Needs Met by High Rate Sedimentation and Filtration," *Public Works*, p. 65 (Jan., 1970).
16. Culbreath, M. C., "Necessary Modifications for High Rate Filtration," *Journal American Water Works Association*, p. 699 (June, 1968).
17. Culp, R. L., "New Water Treatment Methods Serve Richland," *Public Works*, p. 86 (July, 1964).
18. Garel, T. V., "Depth Filters Do Double Duty," *Water and Wastes Engineering*, p. 34 (Dec., 1965).
19. Laughlin, J. E., and Duvall, T. E., "Multiple Media Filters Are Key to Texas Plant Expansion," *Water and Wastes Engineering*, p. 76 (May, 1968).
20. Spink, C. M., and Monscvitz, J. T., "Design and Operation of a 200 mgd Direct Filtration Facility," presented at the 1973 American Water Works Association Conference, Las Vegas, Nevada (May 14, 1973).

Index

Index